The Mini**FARMING**™
Guide to
Fermenting

The Mini**FARMING**™
Guide to
Fermenting

*Self-Sufficiency from Beer and Cheese to
Wine and Vinegar*

Brett L. Markham

Skyhorse Publishing

Skyhorse Publishing books may be purchased in bulk at special discounts for sales promotion, corporate gifts, fund-raising, or educational purposes. Special editions can also be created to specifications. For details, contact the Special Sales Department, Skyhorse Publishing, 307 West 36th Street, 11th Floor, New York, NY 10018 or info@skyhorsepublishing.com.

Skyhorse® and Skyhorse Publishing® are registered trademarks of Skyhorse Publishing, Inc.®, a Delaware corporation.

www.skyhorsepublishing.com

10 9 8 7 6 5 4 3 2 1

Library of Congress Cataloging-in-Publication Data is available on file.

ISBN: 978-1-61608-613-8

Printed in China

Contents

Acknowledgments

This book draws on a lot of chemistry knowledge. Though I had chemistry instructors in college, the groundwork was laid much earlier by my father's encouragement in providing me chemistry materials as a child even after I accidentally gassed myself. Two teachers who were very inspiring to my pursuit of science were Mr. Booth, my eighth grade Physical Science teacher who provided demonstrations of exploding sodium (nothing entices a boy more than things that go boom!), and Mrs. Firestone, my teacher for Chemistry I & II in high school, whose love and zeal for the subject was infectious.

A book is sort of like sausage. You see the finished product but you don't necessarily notice everything that went into it. The primary ingredient in a book is time. When you already work a full-time job, that time comes at the expense of other people in your life. I would like to explicitly thank my wife Francine for her editorial assistance and my daughter Hannah for her help with pictures. I also appreciate the patient indulgence of my family and many friends who endured a long leave of absence so I could write books I felt would be helpful.

There were also a lot of experiments along the way. I'd like to thank many people I value across the country for their support and for often sampling my concoctions and giving their honest opinions, including my mother-in-law, co-workers at BELD and friends across the country. I kept telling them if they drank enough of my wine it would start to taste good. Eventually … it did.

PART I

Introductory Considerations

1

Introduction to Fermented Foods

Though it may be hard to imagine, it is only in comparatively recent times that humans have had access to technology such as freezers, refrigerators, and canning as a means of preserving food. Even today, many parts of the world have scant access to electricity or modern devices. In these areas, as well as in historic times in locales where the availability of food has varied with the seasons, preservation of food for lean times was and still is extremely important. Likewise, in many periods of history, the knowledge of sanitation has been poor, and the water supply a source of illness. Hence, fermented beverages free of disease-causing bacteria and protozoans were more commonly consumed than water.

Fermentation is one of the oldest means of food preservation known to humanity, perhaps surpassed in age only by dehydration. The underlying idea of using fermentation to preserve food is straightforward: Some element of the food that would ordinarily support rotting is converted into something that retards rotting. In this fashion, the shelf-life of the food is increased while retaining many of its beneficial nutritional qualities.

In the case of wine or beer, sugars that would support many disease organisms are converted into alcohol, which is a potent germicide. In the instance of yogurt and similar soft cheeses, the sugar lactose is converted to lactic acid, which lowers the pH of the milk sufficiently that many rotting organisms cannot survive. Vinegar is a further oxygenation of alcohol that is both directly toxic to many microbes but, like lactic acid, it also lowers the pH to such a level as to create an environment hostile to the survival of disease organisms.

Sometimes fermentation is simply used to make something that is more tasty or portable. Certainly, cheese is more portable than milk, and bread is more tasty than cracked wheat kernels.

Fermented foods have progressed well beyond their start as a means of food preservation, and are now a cornerstone of the human diet. Fermented foods are not just made to preserve their ingredients: they are culinary elements in their own right. I sincerely doubt that most people think of wine as a means of preserving the antioxidants of grape juice or of cheese as a means of preserving the proteins in milk for long shelf life. Instead, these are produced and consumed as dietary constituents that are, in many cases, preferred over their initial ingredients.

Fermented foods cost money. Because of the additional work required to create them, they usually cost considerably more than their raw ingredients would cost. Furthermore, some fermented foods have costs added in the form of "sin taxes" that make them even more expensive. Sometimes, it is difficult if not impossible to even buy certain fermented foods at all, such as wine made from organic apples and local honey. I have found local wines that include honey, but at $32/bottle they are cost-prohibitive. Making your own will give an excellent product at a much more reasonable price.

When I set out to write this book, I was already broadly familiar with the many books in print addressing these subjects. Some of them are truly excellent and charmingly written. I treasure them and enjoy them, and hope you will too. But almost all of these books are predominantly dedicated to recipes. They provide a bit of explanation, but their primary thrust is showing you how to duplicate what someone else has already done. Though this can be useful, it isn't true self-sufficiency. What if someone gives you a bushel of starfruit that you want to turn into wine but you can't find a recipe? How do you make your own malt for beer?

Many other books are essentially chemistry books. As a man with his own chemistry lab, I can really appreciate a good chemistry book! The problem is that most people haven't taken chemistry since high school and are somewhat stymied by the chemical equations and their meanings. A discussion of pH in terms of logarithmic concentration of H+ ions may be entirely factual, but also perfectly useless. Even though these chemistry books definitely contain all of the information needed to adapt recipes, translating them into something actionable by a non-chemist in the real world can be challenging.

In my earlier books, *Mini Farming: Self-Sufficiency on ¼ Acre* and *Maximizing Your Mini Farm,* I endeavored to convey the underlying principles and reasoning for the methods rather than just instructions. That is my objective with this book as well. I want to provide sufficient background in the underlying chemistry to enable you to be flexible and adaptive, but concentrate on practical nuts-and-bolts skills needed to convey self-sufficiency quickly.

Don't get me wrong. This sort of material can only be simplified so much before it becomes useless, so I am going to cover the chemistry and some math when needed. But after covering those, I am going to compress that information into certain ideas and principles that will allow you to use them easily.

In some ways, this book will seem a bit more complex than its predecessors, but all of the knowledge will be built hierarchically from a sound foundation so as to be easily understood if you don't skip chapters. I'll keep things conceptual rather than turning this into a text requiring memorization of Avogadro's Number.[1] Knowledge is power, and knowledge that lends itself to greater self-sufficiency is well worth acquiring.

---

1   Avogadro's Number is the number of atoms or molecules of a substance contained in a number of grams of that substance equal to its atomic or molecular weight.

2

Basic Chemistry for Fermentation

There are some basics of chemistry you'll need to know for the following chapters and for the rest of this book to make sense. Most of this, after second year high school chemistry, should just be a review. But just in case it has been a while since high school, I'm going to review enough information so that what follows will be straightforward.

Conservation of Matter

Matter cannot be created or destroyed. It can change form, but cannot be made to simply appear or disappear.[2] As a result,

---

2   Strictly speaking, as Einstein revealed, energy is matter and matter is energy, so matter can in fact be converted to energy and vice versa. But for our purposes, since we won't be using nuclear reactions in the kitchen, this physical law can be expressed simply as "Matter can neither be created nor destroyed."

any physical, biological, or chemical manipulation of a substance will always result in components having the same total mass as the starting material. In addition, though transmutation of elements is indeed possible in a nuclear reactor, for all ordinary chemical and biological processes, one fundamental element cannot be turned into another. In other words, carbon cannot be turned into silicon and lead cannot be turned into gold. So all of the fundamental elements involved in a physical, chemical, or biological process will also be present in the result of that process, and in the same quantities.

Solutions versus Mixtures

A solution is a solvent, such as water, in which some other substance, such as sugar or salt, has been dissolved. In a solution, the concentration of the dissolved substance is constant throughout the solvent. The opposite of this is a simple physical mixture, such as vinegar and oil dressing. Neither is dissolved in the other, and they can be easily separated.

Elements, Compounds, and Chemical Formulas

Substances can be divided into two basic categories: elements and compounds. Elements are the fundamental building blocks of ordinary matter. Under ordinary circumstances, outside of a nuclear reactor or a particle accelerator, elements cannot be broken down into any more basic constituents. Examples of elements are copper, chlorine, oxygen, and nitrogen. The periodic table of the elements contains a listing of all known elements along with their most important chemical properties. Each element has a symbol to represent it. For example, carbon is represented as C, hydrogen as H, and oxygen as O. Some elements have symbols that hearken back to German or Latin names for the elements. For example, lead is represented by Pb, which is short for plumbum (from whence the word plumber is also derived) and sodium is represented by Na, which is short for natrium. These symbols are used in chemical formulas to represent those elements.

Compounds are substances comprised of more than one element. For example, table salt is a compound comprised of sodium(Na) and chlorine(Cl), and a molecule of salt is represented by the formula NaCl. Some compounds are more complex and have more than one unit of a given element. In such cases, the proportion of that element is noted in the formula as a subscript. For example, a

Periodic Table of Elements

1 1A	2 2A	3 3B	4 4B	5 5B	6 6B	7 7B	8	9 8B	10	11 1B	12 2B	13 3A	14 4A	15 5A	16 6A	17 7A	18 8A
																	2 He Helium 4.00
	4 Be Beryllium 9.01											5 B Boron 10.81	6 C Carbon 12.01	7 N Nitrogen 14.01	8 O Oxygen 16.00	9 F Fluorine 19.00	10 Ne Neon 20.18
	12 Mg Magnesium 24.31											13 Al Aluminum 26.98	14 Si Silicon 28.09	15 P Phosphorus 30.97	16 S Sulfur 32.07	17 Cl Chlorine 35.45	18 Ar Argon 39.95
	20 Ca Calcium 40.08	21 Sc Scandium 44.96	22 Ti Titanium 47.87	23 V Vanadium 50.94	24 Cr Chromium 52.00	25 Mn Manganese 54.94	26 Fe Iron 55.85	27 Co Cobalt 58.93	28 Ni Nickel 58.69	29 Cu Copper 63.55	30 Zn Zinc 65.39	31 Ga Gallium 69.72	32 Ge Germanium 72.61	33 As Arsenic 74.92	34 Se Selenium 78.96	35 Br Bromine 79.90	36 Kr Krypton 83.80
	38 Sr Strontium 87.62	39 Y Yttrium 88.91	40 Zr Zirconium 91.22	41 Nb Niobium 92.91	42 Mo Molybdenum 95.94	43 Tc Technetium (98)	44 Ru Ruthenium 101.07	45 Rh Rhodium 102.91	46 Pd Palladium 106.42	47 Ag Silver 107.87	48 Cd Cadmium 112.41	49 In Indium 114.82	50 Sn Tin 118.71	51 Sb Antimony 121.76	52 Te Tellurium 127.60	53 I Iodine 126.90	54 Xe Xenon 131.29
	56 Ba Barium 137.33	57 La Lanthanum 138.91	72 Hf Hafnium 178.49	73 Ta Tantalum 180.95	74 W Tungsten 183.84	75 Re Rhenium 186.21	76 Os Osmium 190.23	77 Ir Iridium 192.22	78 Pt Platinum 195.08	79 Au Gold 196.97	80 Hg Mercury 200.59	81 Tl Thallium 204.38	82 Pb Lead 207.2	83 Bi Bismuth 208.98	84 Po Polonium (209)	85 At Astatine (210)	86 Rn Radon (222)
	88 Ra Radium (226)	89 Ac Actinium (227)	104 Rf Rutherfordium (261)	105 Db Dubnium (262)	106 Sg Seaborgium (266)	107 Bh Bohrium (264)	108 Hs Hassium (269)	109 Mt Meitnerium (268)									

58 Ce Cerium 140.12	59 Pr Praseodymium 140.91	60 Nd Neodymium 144.24	61 Pm Promethium (145)	62 Sm Samarium 150.36	63 Eu Europium 151.96	64 Gd Gadolinium 157.25	65 Tb Terbium 158.93	66 Dy Dysprosium 162.50	67 Ho Holmium 164.93	68 Er Erbium 167.26	69 Tm Thulium 168.93	70 Yb Ytterbium 173.04	71 Lu Lutetium 174.97
90 Th Thorium 232.04	91 Pa Protactinium 231.04	92 U Uranium 238.03	93 Np Neptunium (237)	94 Pu Plutonium (244)	95 Am Americium (243)	96 Cm Curium (247)	97 Bk Berkelium (247)	98 Cf Californium (251)	99 Es Einsteinium (252)	100 Fm Fermium (257)	101 Md Mendelevium (258)	102 No Nobelium (259)	103 Lr Lawrencium (262)

❯ Periodic Table of Elements

molecule of glucose is represented as $C_6H_{12}O_6$, meaning that it contains six atoms of carbon, six atoms of oxygen and twelve atoms of hydrogen. This is what is called an empirical formula. It tells you how many atoms of which elements are contained in the compound, but nothing about its structure.

Empirical and Structural Formulas

Sometimes the formula for a compound is written in such a way as to give some information about its structure in space. That is, the formula gives some indication as to which elements are connected to each other or that give clues about the nature of the compound. For example, ethyl alcohol might be represented by the structural formula CH_3CH_2OH rather than the empirical formula C_2H_6O. The structural formula represents the following arrangement of elements:

By writing the formula as CH_3CH_2OH a person familiar with chemistry can instantly tell from the -OH ending that

```
    H   H
    |   |
H - C - C -OH
    |   |
    H   H
```

❯❯ Ethyl Alcohol

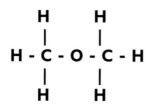

the compound is an alcohol. If it had instead been written as CH_3OCH_3, this compound containing the same elements in the same proportions, but with a different structure, would be dimethyl ether, a dramatically different substance used to start cars and in prior decades as an anesthetic. Clearly, the elements and proportions of a compound are important, but the structure is equally so.

Atomic and Molecular Weights

In the included Periodic Table of Elements, you'll find that each element is listed with an atomic weight. (It's the number near the bottom in each box.)

Atomic weight can be explained in various ways. But for our purposes in chemical formulas, the atomic weight of an element is the number of grams of that element needed to have a certain number of atoms. That number is Avogadro's Constant, 6.0221415×1023. This quantity of atoms is known as a mole. You don't need to memorize this number. The important thing to know is that one mole of an element will have a mass in grams equal to its atomic weight. So by looking at the periodic table, we can readily see that one mole of sulfur will have a mass of 32.06g while a mole of calcium will have a mass of 40.078g.

Compounds have what is called molecular weight or formula weight. The molecular weight of a compound is computed by adding the atomic weights from the periodic table of each constituent atom. The atomic weights of carbon, hydrogen, and oxygen respectively are 12.0107, 1.00794 and 15.9994. So the molecular weight of glucose, $C_6H_{12}O_6$, would be 6 x 12.0107 + 12 x 1.00794 + 6 x 15.999 or 180.156. This means that one mole of glucose, that is, a number of atoms equal to Avogadro's Constant, will have a mass of 180.156 grams.

Normality

Don't worry, we're not delving into psychology! Normality in chemistry refers to how much of a solute has been dissolved in a solvent, and is equal to moles per liter. The unit for normality is an uppercase N. We already know that one mole of glucose has a mass of 180.156 grams, so we could make some 1 N glucose solution by putting 180.156 grams of glucose in a beaker, and then adding enough distilled water to fill the beaker to the one liter line. If we wanted a 0.1 N solution, we'd multiply 180.156 grams by 0.1, to get 18.016 grams.

The primary reagent we will be using in such a concentration is sodium hydroxide. We will be using 0.1 N sodium hydroxide as a measurement standard to determine the amount of acid present in wine. Using the molecular mass of sodium hydroxide of 39.997, you could make up the reagent yourself, then, by adding 3.999 grams of sodium hydroxide to a beaker, and then adding enough distilled water to make a liter.

In practice, the markings on beakers aren't accurate enough for such work, and you'd use a volumetric flask or a graduated cylinder instead. And as one liter is way more of this than we'll ever use, we'd be more likely to cut everything in quarters like a recipe, and add just one gram of sodium hydroxide and enough water to make up 250 ml.

Sodium hydroxide solution is widely available from wine suppliers in 0.2N concentration. For our procedures calling for 0.1N concentration, you can create 10ml of 0.1N solution by adding 5 ml of 0.2N sodium hydroxide and 5 ml of distilled water to a 10 ml graduated cylinder.

Density and Specific Gravity

Every substance has a certain density. Density is measured as the mass of the substance in grams that occupy one cubic centimeter. The density of water is 1.0 grams per cubic centimeter, and the density of glucose is 1.54 grams per cubic centimeter, abbreviated as 1.54 g/cm^3. The density of ethyl alcohol is .789 g/cm^3. Substances heavier than water will have a density greater than one (1), and those that are lighter than water will have a density of less than one (1).

Density can be measured in various ways. A classic method when dealing with solids such as iron is to weigh a sample of the substance, and then immerse that substance in water and measure how much volume it displaces. When the mass is divided by the volume, you get the density. You could do the same with a liquid by weighing an empty graduated cylinder, filling the cylinder to a certain volume, then weighing the filled cylinder. If you subtract the empty weight from the filled weight and divide by the volume, you'll get the density.

In practice, when making beer and wine, an alternative measure of density is used, called Specific Gravity. Specific gravity is measured using a hydrometer. A hydrometer is a calibrated weighted device that looks like a thermometer with lead shot in the bottom. The less dense the liquid, the deeper the hydrometer sinks and vice versa. Because the density of a substance varies with temperature, hydrometers are calibrated at a given temperature. If the density is being measured at a temperature substantially different than the temperature used to calibrate the device, a compensation factor is added or subtracted.

pH

pH is a measurement of how acidic (like lemon juice) or alkaline (like ammonia) a solution is. Distilled water is neutral, and has a pH of seven. Alkaline solutions have a pH greater than seven, and acidic solutions have a pH less than seven. The underlying science of pH is relatively complex. The number represents the negative base ten logarithm of the molal concentration of hydronium (H_3O+) ions in the solution. For our purposes, though, it is most important to understand that the pH of a solution affects what microorganisms can live in it, and also how chemical reactions proceed. For example, when making beer, a great deal of tannin will be extracted from grain with boiling water at a pH near seven, but that very little tannin will be extracted in a hot mash with a pH of less than five.

pH can be measured using test papers, indicator solutions such a phenolphthalein, or using a meter. In past books I have described how to test pH using phenolphthalein, but as the accuracy of using a meter is greater and the price of meters has declined, I now recommend getting an inexpensive pH meter.

Organic Chemistry

When I speak of organic chemistry, the term "organic" has no relation to its meaning in the context of organic gardening. Instead, it is referring to compounds that contain carbon. Some high school and college chemistry teachers may give a collective gasp at how quickly I am going to summarize this topic because I will be skipping the systematic naming systems and a discussion of covalent bonding. But my goal is to convey the information needed to deal successfully with fermented foods and understand what is going on in broad terms rather than pass the GRE Chemistry exam.

Carbon-based organic compounds lie at the very root of life. Whether it is the amino acids that form the DNA in the nucleus of cells or the esters that provide many of the unique tastes and smells of food, all of these compounds are carbon based. Carbon atoms have a unique ability to bond with other elements and also to form chains of theoretically unlimited length by bonding with other carbon atoms. Silicon can form similar bonds that are the basis for silicone products, but with nowhere near the flexibility of carbon. In essence, no carbon = no life.

The sugars, alcohols, acids, proteins, tannins, and carbohydrates present in the materials we use to make fermented foods are all organic compounds, and they don't just sit there and do nothing. Given time, the right conditions, and the catalytic action of enzymes generated by microorganisms, these compounds transform and combine to make new compounds.

At first blush, the idea that wine would taste like anything other than grape juice in which the sugar has been replaced with alcohol seems preposterous. But on closer examination, it is discovered that the organic acids present in the wine will combine with the alcohols present to create new compounds with altogether new flavors that weren't present in the original juice, and the tannins that are present will combine with the proteins to form insoluble compounds. So put on your seat belt for a whirlwind tour of the basics of organic compounds.

Alcohols

When the term "alcohol" is used, we are usually speaking of ethyl alcohol, which is the type produced by fermentation and distillation. But rubbing alcohol is an altogether different compound described as isopropyl alcohol and the so-called "dry gas" added to gas tanks is yet another type of alcohol known as methyl alcohol.

There are many different types of alcohols. In general, you can tell that something is an alcohol in a chemical sense by the last two letters of its name. If it ends with -ol it is an alcohol. Examples include glycerol (a core constituent of fatty acids), propylene glycol (used in foods), ethylene glycol (a poisonous compound used as antifreeze), and so forth. Alcohols are distinguished structurally by the presence of an -OH group attached to a carbon atom. The structural formula for methanol (AKA methyl alcohol) is CH_3OH, and the formula for ethanol is CH_3CH_2OH. This series proceeds indefinitely, and there are many variations on the theme.

Alcohols with more than two carbon atoms are referred to as "higher alcohols" within the context of fermentation. To some extent, in small quantities, these are naturally present in some foods. Others will be produced in small quantities by yeast and bacteria as part of the natural process of fermentation.

The important thing that needs to be understood is that an alcoholic fermentation, though presented as though ethyl alcohol is the only result, will actually result in a wide array of alcohols being produced in very small quantities. These other alcohols will affect the taste directly, and they will also combine with other constituents to form altogether new compounds that will create new flavors and smells. The temperature, specific strain of yeast and the other nutrients present all have an effect on the proportion of higher alcohols produced. This is why two otherwise identical batches of wine will taste differently if their respective fermentation and aging were carried out at different temperatures.

If you have heard tales of people going blind from drinking moonshine, the reason is because it was distilled improperly. Different alcohols boil at different temperatures. Methyl alcohol, which turns into formate in the body and damages the delicate optic nerves, is produced along with ethyl alcohol in the yeasty mash that is distilled to make moonshine. Methyl alcohol evaporates at a lower temperature than the ethyl alcohol people drink. When moonshine is distilled improperly, the methyl alcohol is collected along with the ethyl, and this is what makes it dangerous.

Though we won't be discussing distillation of liquors in this book, I think it is worthwhile to know that different types of distilled liquors such as whiskey, rum, brandy, and so forth have different tastes—not only because of what was used as a starting material—but because when they are distilled a different proportion of higher alcohols, esters, and other compounds are passed into the final product because of differences in time, temperature, and so forth. In all of those cases, the deadly methyl alcohol is allowed to pass off without being collected before the actual portion intended for human consumption is collected from the still.

Just as liquors have different flavors and odors due to these different proportions of alcohols, the same can occur with fermented products that have not been distilled. The variety of yeast used, the starting ingredients, the temperature of fermentation, and the nutrients that are present all have an effect on the proportions of higher alcohols produced.

Organic Acids

Organic acids are a part of our everyday lives and are constituents of every living thing. Uric acid is present as a waste product in our urine, citric acid is commonly found in citrus fruits and tartaric acid is the primary acid in wine grapes. But this just scratches the surface. Upon close examination it is discovered that nearly all foods that we will use in fermentation contain a vast array of organic acids that impart flavors and smells of their own. In chemistry, acids

» The structure of propanoic acid.

can be readily identified by structural formulas ending in -COOH. For example, butyric acid (found in butter and cheeses) has the structural formula $CH_3CH_2CH_2$-COOH, and propanoic acid (present in Swiss cheese) has the structural formula CH_3CH_2COOH.

Organic acids also play a key role, along with alcohols, in the creation of new flavors in fermented foods, particularly if those foods are aged. Acids and alcohols combine to form esters.

Esters

Esters are naturally present in fruits and vegetables, and are fundamental in providing their unique and distinguishing flavors and scents. Esters are also formed, albeit slowly, by the combination of acids and alcohols. That is to say that over time, the acids and alcohols in fermented foods will

⊗ The structure of ethyl acetate.

combine to create flavoring esters that weren't there in the first place, and thereby develop unique depth and complexity.

For example, small quantities of acetic acid, CH_3-COOH, are present in wine, as is ethyl alcohol, CH_3CH_2OH. When these combine, a molecule of water is lost, and ethyl acetate, CH_3-COO-CH_2CH_3, is created. Ethyl acetate has a taste and smell similar to bananas. Because humans can detect tastes and smells of compounds with concentrations measured in parts-per-million, the fact that new esters are produced in very small quantities doesn't mean they can't make a difference in the taste.

Aldehydes and Ketones

The complete oxidation of alcohols (through burning for example) results in water and carbon dioxide as the final products, but several intermediate states of oxidation are possible. Aldehydes and ketones are created through the partial oxidation of alcohols.

The most common aldehydes used in industry are formaldehyde and acetaldehyde, but many other aldehydes occur in nature, including vanillin and cinnamaldehyde. (You can probably guess the flavors these impart just from the names!) An aldehyde can be distinguished in chemistry by the presence of a formyl group, designated -CHO. Thus, benzaldehyde, which gives almonds their characteristic odor, has the chemical formula of C_6H_5CHO.

Aldehydes are very reactive, and it is very easy for them to be further oxidized, thus changing them to yet another compound. This is one reason why fermented foods are protected from air and why those, such as sherries that have greater air exposure, develop defining tastes and smells.

Ketones are very similar to aldehydes, except that they form from what are called "secondary" and "tertiary" alcohols. A secondary alcohol has the defining -OH group attached to a carbon atom with only one other hydrogen atom attached, and a tertiary alcohol has the -OH group attached to a carbon with no hydrogen atoms attached.

In the case of secondary alcohols, the addition of an oxygen atom (i.e. oxidation) steals the one H from the carbon along with the H from the -OH group to create water, and the carbon is left with only an oxygen attached. The defining structural group of a ketone, then, is an oxygen double-bonded to a carbon: C=O. As an example, tertiary butyl alcohol may be partially oxidized to acetone, also known as methyl-methyl ketone (CH_3-CO-CH_3). Ketones are substantially less reactive than aldehydes, and occur pervasively in nature. Sugars (such as fructose) are a form of ketone, collectively referred to as "ketoses."

Proteins and Amino Acids

Proteins are formed from chains of amino acids, making amino acids the fundamental building blocks of life. Their structure is similar to that of organic acids in that they have a defining -COOH group, but they also have an -NH_2 group replacing the hydrogen on the adjacent carbon atom.[3] The simplest amino acid is glycine, whose formula is CH_2NH_2COOH, but there are many others. Some amino acids are deemed essential to an organism if that organism cannot synthesize them from other amino acids. In humans, only nine amino acids are essential. Given an adequate diet, humans can synthesize whatever alanine we need, but we can't synthesize tryptophan or methionine and would ultimately starve to death

---

3 The amino acids I am referencing in this case are alpha amino acids. There are also branched-chain amino acids, gamma amino acids, and so forth.

for their lack no matter what else we ate. Every organism is somewhat different in the amino acids it can or cannot synthesize.

Because of the large number of amino acids and the fact that they can form chains in any order and that the chains can be of any length, the number of proteins that can be created in this fashion is essentially unlimited and accounts for the amazing complexity of life. Proteins and amino acids are present in all plant and vegetable matter.

Chemistry Equipment

In this book I will be using a chemistry approach to certain things, and it will be easier if you have certain basic laboratory equipment on hand. This equipment will last your lifetime, isn't very expensive, and will help make your work faster, easier, and more accurate. The precise glassware and chemicals needed will be noted in the description of each procedure, but at a bare minimum you will need a pH meter and a digital scale of 100g or greater capacity that is accurate to +/- .01g and a calibration weight for digital scale.

PART II
Making Wine

3

Overview of Winemaking

Anyone who has seen a wine critic on television can be a bit intimidated by the prospect of trying to make a wine that is even drinkable, much less enjoyable. Fortunately, Frederic Brochet conducted two studies using 57 wine experts at the University of Bordeaux in 2001 that will forever put the wine experts into perspective.[4]

In the first study, the experts were given two glasses of wine to describe, one being a white wine and the other a red wine. Unknown to the experts, both glasses were a white wine but the wine in one of the glasses had been dyed red. Not even one of the 57 experts at the University of Bordeaux could distinguish that the red wine was really white, and they even went on to

4   Downey, R. (2002), Wine Snob Scandal, *Seattle Weekly*, Feb. 20, 2002

describe the fake red wine as having characteristics associated with red wines such as "tannic notes."

In the second study, a cheap wine was put into bottles denoting both a cheap and an expensive wine. Same wine, different bottles. The experts described the wine in the expensive bottles as "woody, complex, and round" while describing the exact same wine in the cheap bottles as "short, light, and faulty."

What this means is that you need not be intimidated by wine snobbery. All you need to do is make a sound product using good ingredients and proper methods, and as long as you put it in a nice bottle with a nice label and serve it in a nice glass it will be fully appreciated. Perception of the details really matters in the impression, so even if you performed your primary fermentation in a plastic bucket, don't you dare serve it in a plastic cup!

Winemaking is among the oldest methods of food preservation. In wine the levels of sugar in the original juice are reduced to make the juice unattractive to organisms that require sugar for growth, and the sugar is replaced with alcohol that makes the juice an inhospitable environment for most spoilage organisms. Meanwhile, many of the beneficial nutrients in the original juice are preserved, including vitamins and antioxidants. The yeast used to convert the sugar to alcohol also imparts a number of B vitamins to the mix.

Later, wine became an end in itself for which fruits were grown, and an entire culture and mythology have grown up around grapes, winemaking, and wine. What started as a method of preserving the essential nutrients of grapes and compensating for dangerous water supplies in an era when aseptic packaging and refrigeration did not exist has now grown into a multi-billion dollar global industry, and there are bottles of wine that can only be had for a cost exceeding that of a new car.

The term wine, in the purest sense of the term, applies only to the results of fermenting the juice of European vitis vinifera grapes. These are a species that is distinct from the grapes indigenous to North America, and only vitis vinifera grapes—and no other grape or fruit—have the right levels of sugar, tannin, acidity, and nutrients to produce wine without adding anything. Grapes grown in particular regions lend their unique flavors to wines named after them, such as Champagne; and wines produced by some wineries have even become status symbols, such as those produced by Château Lafite Rothschild.

Here, however, I am using the term "wine" to refer to country wines. That is, wines made from any fruit available and to which sugars, tannins, organic acids, spices, and other ingredients have been added to not only compensate for the areas in which the ingredients fall short of vitis vinifera grapes, but also to create their own experience of taste and smell.

Country wines are in no way inferior, and in fact being free of the constraints of traditional winemaking leaves you open to experiment broadly and create delightfully unique wines that are forever beyond the reach of traditional wineries. Home winemaking of up to 200 gallons annually is legal in the United States so long as you don't try to sell it. (If you try to sell it you will run afoul of the infamous "revenuers" who always get their culprit in the end.)

Americans drink nearly two gallons of wine per capita annually, meaning the consumption for a standard household is just shy of eight gallons. That's 40 bottles of wine. If you figure $12/bottle, that's nearly $500 per year. At that rate, and using fruit you either get inexpensively in season or free in your back yard, you will quickly recoup your investment in materials and supplies. In addition, homemade wines make excellent personal gifts; every year I give bottles to friends and business associates for holiday presents. The wine is always appreciated.

The technique of making wine is conceptually straightforward. The juice (and sometimes solids) of a fruit are purified of stray microbes and supplemented with sugar, acid, and other nutrients to make a must, inoculated with an appropriate strain of yeast, and fermented. The fermentation takes place in two distinct phases. The first phase, known as the primary fermentation, is very fast and lasts only a couple of weeks. The wine is then siphoned (a process called racking) into a new vessel and fitted with an airlock where it may continue its secondary fermentation for several months with a racking again after the first couple of months or any time substantial sediment has formed.

I'll explain the specifics of the equipment throughout this chapter, along with some of the nuances. For now, I want to convey that if you normally consume wine or would give wine as a gift, making your own quality wines is inexpensive, fun, and easier than most would believe.

There are places where you will be tempted to skimp or make do on the list of ingredients and equipment I am about to present, but let me encourage you to get everything on the list. Because the techniques I describe rely on natural ingredients whose constituents will vary, all of the testing equipment is necessary. The other equipment and ingredients are necessary to maintain sanitary conditions or make a quality wine. If you don't live near a store for wine hobbyists, there are a number of excellent sources on the Internet that can be located via a web search. To save on shipping, I'd recommend getting as much from one store as possible.

Winemaking Equipment
Primary Fermenter

The primary fermenter is a large plastic bucket made of food-grade plastic. It is sized at least 20% larger than the largest batch of wine you plan to make in order to keep the constituents of the vigorous primary fermentation from spilling out of the fermenter and making a mess you will not soon forget. The bucket should be equipped with a lid and gasket, and also have provisions for fitting an airlock. These are available in various sizes from beer and wine hobby suppliers. I'd recommend a two-gallon and a six-gallon bucket. Even though it is possible to get these buckets for free from restaurants, I would advise against it as most were used to hold something that was previously been pickled using vinegar. You don't want vinegar organisms in your wine.

In general, the primary fermentation evolves carbon dioxide so rapidly that an airlock isn't strictly necessary. Furthermore, the first stages of fermentation require oxygen until the yeast cells multiply enough to reach a critical mass before the start of fermentation. Just plugging the hole in the lid with a clean cotton ball that allows air movement but blocks dirt, dust, and insects will suffice. (Replace the cotton ball if it becomes saturated with must.) Even so, I usually use an airlock after the first week.

It is possible for the smells and tastes of plastic to become infused into wine. This is not a concern for the primary fermentation because the wine is only in contact with the container for a couple of weeks. Also, you will have used a container of food grade plastic selected for its low diffusion which you cleaned thoroughly prior to use.

Secondary Fermenter

Because the wine stays in contact with the secondary fermenter for months or even years, this is best made of glass. You can also use specially made oak casks for long secondary fermentation, but these are very expensive and need special care and maintenance. So for now, I would skip the oak casks.

» One- and five-gallon primary and secondary fermenters.

The glass vessels come in various sizes from one gallon up to five gallons. The smaller one-gallon vessels are just one-gallon jugs, and the larger three- or five-gallon vessels are glass carboys used on water coolers.

You will also see some plastic carboys available in winemaking magazines and from various suppliers. These are advertised as being made in a way that makes them impervious to the diffusion of the plastic into the wine, and they offer the advantage of being much lighter than glass so the shipping costs are lower. Nevertheless, plastic is harder to clean than glass, so I would not recommend these if glass can be obtained instead.

You will also need to get a special brush for cleaning your jug or carboy because the opening is too small for even the smallest hands and a regular bottle brush is too short and isn't bent for cleaning around the edges.

The fermentation that takes place in the secondary fermenter is long and slow. As the carbon dioxide is evolved more slowly, it is possible for air to be drawn into the vessel, especially if temperatures change. During secondary fermentation, you want to prevent oxygen from coming into contact with the wine, because oxygen adversely affects the quality of the wine by changing the character of some of the evolved organic compounds.

By fitting the hole in the fermenter with a stopper and an airlock, you will allow a protective blanket of carbon dioxide to cover the surface of your wine. You will need rubber stoppers with one hole in them that are sized correctly for your secondary fermenter. The airlock is prepared, put into the hole in the stopper, and then the stopper is placed in the hole at the top of the fermenter.

Because you will be racking your wine from one secondary fermenter into another, you need two secondary fermentation vessels.

One thing that people often overlook is a carrying handle. If you are making wine in batches larger than a gallon, those carboys are extremely heavy and difficult to handle. The handle that you order can be installed on a carboy and then removed to be used on another, so you only need one. They cost about $10 at the time of this writing and are well worth it as they make the task of handling carboys a great deal easier.

Airlocks

Airlocks are devices installed on a fermenter that allow gas to escape, but do not allow air to leak back in. They come in a variety of configurations, but all are filled with water or a solution of potassium metabisulfite. The airlock is filled to

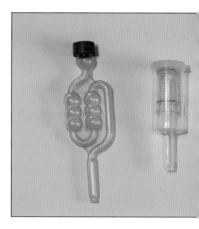

the level specified on the device, inserted in a one-hole rubber stopper and then attached to the fermenter. You should have at least two of these. The style you choose doesn't usually matter, but if your wine will experience swings in temperature, avoid the type illustrated on the left because the liquid in it could be sucked back into the fermenter.

Racking Tube

A racking tube is a long two-part tube that is inserted into the wine and pumped to start a siphoning action in order to transfer the wine from one container into another. It has a knob at the bottom that directs the flow of fluid in such a way as to minimize the amount of sediment transferred in the process. You will also need five feet of plastic tubing to go with it. A stop-cock, which is a plastic clip that can be used to stop the flow temporarily, will come in handy when using the racking tube to transfer wine into final wine bottles for corking.

Racking tubes come in two sizes; one that is smaller and will fit into a gallon jug and one that is larger and will not. Get the smaller one initially as it will work for both gallon jugs and five-gallon carboys.

Always clean your racking tube and plastic tubing before and after use, and run a gallon of sulphite solution through it to sterilize the components. Otherwise, it will accumulate debris attractive to fruit flies that carry vinegar bacteria and you will unwittingly start manufacturing vinegar instead of wine. The tubing is inexpensive and it is best to replace it after several uses.

Corker

Corkers are used to insert corks into wine bottles. As you've discovered if you have ever tried to put a cork back into a wine bottle, corks are slightly larger than the holes they are

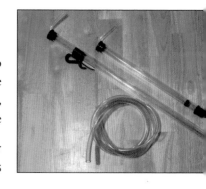

» Vinyl tubing and racking tubes sized for five-and one-gallon fermenters

❯❯ I've had the corker on the left for eight years. The one on the right is just a toy.

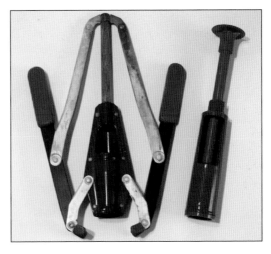

intended to fit. A corker compresses the cork enough for it to slip inside the bottle. Corkers come in many sizes and styles, but I would recommend a metal two-armed lever model which, although somewhat more expensive than the plastic models, does a better job and will serve you well for your lifetime.

Corks need to be soaked before insertion. You don't want to inadvertently transfer a spoilage organism on the corks, so I recommend boiling the corks for 20 minutes and then allowing them to set in the boiled water for another 10 minutes before corking. This will make the corks pliable without contaminating them.

Another problem you may encounter is the cork backing out of the bottle after it was inserted. This is caused by the fact that the cork fits so tightly that the air in the bottle is compressed as it is inserted. The compressed air forces the cork back out of the bottle. You can solve this problem with a bent sterilized paper clip. Straighten the paper clip except for a hook that you leave for it to hang on the edge of the bottle's mouth. Insert the straight part into the bottle mouth and leave it hooked in the edge. Insert the cork as usual. The paper clip has allowed room for compressed gases to escape. Pull the paper clip out and you are done.

Wine Thief

A wine thief is a long tube with a special valve on the end that allows you to remove wine from a container very easily. Clean and sanitize it before and after use. It is generally recommend that wine removed not be returned to the container to avoid contamination. However, unless you have added an adulterant to the wine (such as sodium hydroxide for testing acidity), as long as the wine thief and any equipment used are cleaned and sanitized, I have never had a problem from putting the wine back into the same container.

The biggest reason why you would want to "steal" wine in this fashion is so it can be tested for specific gravity and acidity. I cover these tests and the required equipment extensively in the next chapters.

Nylon Straining Bags

These are fine-meshed nylon bags with zippered closures used to hold fruit for crushing in a fashion that allows you to remove the solids later with minimal mess. The bags can be cleaned, sterilized, and re-used many times. These come in very handy when making wines from crushed blueberries, cherries, and similar fruits. They come in various sizes in order to accommodate different sized batches of fruits and wine.

Wine Bottles

You will need wine bottles. Usually, light-colored wines are bottled in clear bottles and dark-colored wine in green or brown bottles. This is predominantly a social convention, though the darker glass serves the purpose of protecting the coloring matter in the wine from being bleached out by ultraviolet light and sunshine. Your wine should be stored well away from sunshine anyway.

Either way, you will certainly want to use real wine bottles that require a cork. Real wine bottles usually have a concave section at the bottom that allows for solid sediments to remain separate from the wine and have a top made to facilitate a perfect seal with a cork.

There is debate among experts over the use of plastic, screw-top caps, or genuine cork, and whether this has an effect on the long-term taste and quality of wine. In my opinion corks are best simply because they are easiest. Corks are inexpensive in quantity, easily inserted for a perfect seal using simple equipment, and will literally last forever if a bottle is stored on its side to keep it wet. Unlike the experts, I can't tell the difference between a wine stored in a corked bottle as opposed to one using a screw closure, but I recommend corking because it is easier and cheaper in the long run. Also, it just looks better, and the presentation of your wine is as important as any of its other qualities in terms of the reception it receives.

If you decide to make a sparkling wine, you will need to get bottles specifically for that purpose because ordinary wine bottles aren't rated for that pressure.

You will also need special plastic corks and wire closures that will hold the corks in place on the bottle.

Wine bottles come in 375 ml and 750 ml sizes. You will need five of the 750 ml bottles or ten of the 375 ml bottles for each gallon of wine that you are bottling.

Consolidated Equipment List

The following list will make it easy to get everything you will need for the foreseeable future in one shopping trip. I priced this out with a well-known Internet beer and wine hobby shop for $228.60 plus $63.22 for shipping. At that price for shipping, if you can find the gear locally it is worth the trip. You could also save some money by only getting the equipment needed to make one-gallon batches, and the equipment would only cost $134.75 plus $25.95 for shipping. These costs also don't take into account that it is often easy to get wine bottles for free. I get mine from a co-worker who works part-time at a bar. He brings me a few dozen empty bottles and I give him a couple bottles of wine yearly.

1	Five or six gallon plastic fermenter with sealing plastic lid and grommet
1	Two gallon plastic fermenter with sealing plastic lid and grommet
2	Five gallon secondary fermenters, preferably glass
2	One gallon secondary fermenters, glass jugs
1	Cleaning brush for carboys
1	Carboy handle
2	#6.5 universal rubber stoppers with one hole
2	One-gallon secondary fermenters, glass
2	#6 rubber stoppers with one hole
4	Airlocks

❯ Airlocks, wine thief, racking tube, stoppers, corkers and other gear. These will give many years of faithful service if given proper care.

1	Racking tube, sized to fit the one-gallon secondary fermenters, but will work with both
5 ft	⅜" plastic tubing
1	Hose clamp, ⅜"
1	Wine thief
2	Nylon straining bags
1	Two-handed corker
36	Wine bottles
50	Corks

In addition to equipment, making country wines requires a variety of innocuous but nevertheless important additives. All fruits other than European grapes will require additional sugar in the form of either sugar or honey. Most fruits will lack sufficient acid, though without dilution a few may have too much. Likewise, most won't have sufficient tannin to give a properly wine-like mouth-feel. Of course, yeast will need to be added, and the fruits don't have enough nutrients on their own to sustain a healthy fermentation to completion, so nutrients will need to be added for the yeast.

Citric, Malic, and Tartaric Acids

Though most fruits contain more than one of these acids, citric acid is usually associated with citrus fruits, malic acid with apples, and tartaric acid with grapes. You can buy these mixed together as a so-called "acid blend," but they are inexpensive and I recommend buying them separately. This way, you can use the right acid for the fruit you are using or the character you want your wine to have and you aren't locked in to the formula of a given manufacturer. If you are using a recipe that requires "acid blend" you can make it yourself by thoroughly mixing an equal quantity of each of the three acids together.

The acidity of your must should be checked prior to the beginning of fermentation. Most often, acid will need to be added.

Grape Tannin

Tannins are responsible for the astringent taste of a wine. They are present in the skin and seeds of grapes, and so wines that result from conducting the primary fermentation with the skins and seeds will tend to have more tannin and have more astringency. White wines derived from pressed juice are therefore less astringent than red wines derived from fermenting with the skins.

Ingredients other than grapes can have more or less tannin content, and that content will vary based upon the amount of time whole fruit is left in the primary fermenter as well.

Pectic Enzyme

Pectic enzyme is needed to break down the pectins in fruits so they won't leave a cloudy haze in the wine. Grapes have enough pectic enzyme naturally, but all other fruits you are likely to use will need some help.

Fermentation Inhibitor

It can be difficult to judge when fermentation is completed. Early in wine-making it is also common for the home wine maker to be a bit impatient (and justifiably so!) for the finished product. The unfortunate side effect of bottling a bit too early is a wine bottle with a popped cork (and corresponding mess) or even a shattered bottle. Sometimes you can get lucky and just end up with a barely perceptible sentiment and a lightly sparkling wine. In wine judgings, this is considered a defect in a still wine, but for home use it is a delightful thing. Still, if you want to make sparkling wines it is better to make them on purpose rather than accidentally, because their accidental manufacture is attended by some risk.

Potassium sorbate, a semi-synthetic preservative that inhibits fermentation, is added to wines as a stabilizer to prevent further fermentation. It is used in two instances. First, to absolutely guarantee an end of fermentation in wines that are bottled young. Second, to stop fermentation in wines that are intentionally sweet and the only thing inhibiting the yeast is the high alcohol content.

The positive is that potassium sorbate works well, is generally accepted as safe, and will give you good insurance against exploding bottles. It is seldom noticeable at all when used for young wines and white wines. The downside is that it can develop off-smells in some wines over a period of years. So if you are making a wine that you plan to keep for many years, rather than using potassium sorbate I would recommend bulk aging it for at least a year in a secondary fermenter to assure the end of fermentation prior to bottling.

Another method of ending fermentation is to add supplemental alcohol to the wine in the form of brandy (which is distilled from wine). This process is called fortification. Raising the alcohol level in the wine above the alcohol tolerance of the yeast (usually 20%) assures its dormancy. Fortification is used in the manufacture of port wines. Port wines are typically sweet and dark, though some dry and white ports exist. These sweet wines were stabilized for shipping purposes by racking them into a secondary fermenter that already contained enough brandy (about ¼ of the volume of the wine) to raise the alcohol level to 18% to 20%. This brought about a quick end to fermentation while retaining as much as 10% residual sugar. The stability of port wines can allow them to keep for decades.

Yeast Energizer

Yeast energizer supplies crucial nutrients for yeast that allow it to reproduce and do a good job of converting sugar to alcohol. Any wine made from anything but vitis vinifera grapes will need this. Yeast energizer usually contains food grade ammonium phosphate, magnesium sulfate, yeast hulls to supply lipids, and the entire vitamin B complex, of which thiamine (vitamin B_1) is the most important.

Sulfite

Sulfite is used to retard spoilage organisms and wild yeasts and as an antioxidant. Though it is possible to make sulfite-free wine, its use increases the likelihood of success for beginners, particularly when they are using real fruits instead of pasteurized bottled juices. Sulfite is even permitted in wines labeled as USDA Organic.

You should get two forms of sulfite. The first is potassium metabisulfite in the form of Campden tablets. Campden tablets are sized with the idea in mind of accurate dosing of wine and musts to purify must prior to initiating fermentation and help clear and preserve the wine later. The second is powdered potassium metabisulfite. In powdered form it is used to make sterilizing solutions for sterilizing equipment.

Yeast

Home winemaking has been popular so long across so many countries that there are literally hundreds of varieties of yeast available. Because covering them all would be a prodigious task, I want to cover some common yeasts that will be most generally useful for practically anything you'd like to try to turn into wine. Later, you can branch out and try the other excellent varieties of yeast that are available.

Red Star Pasteur Champagne

This is an excellent all-around yeast for making dry wines. It produces glycerol as well as alcohol, and this gives wines a nice mouth-feel. I particularly like using this yeast in wines containing apple, pear, and flower ingredients because it produces fresh aromas that match these ingredients. It works well at lower temperatures, even as low as 55 degrees, and tolerates up to 16% alcohol.

Red Star Montrachet Yeast

If you don't have much control of the ambient temperature of your must, this yeast is a good choice. It can work at temperatures ranging from 55 to 95 (though it does less well at the extremes than it does in the middle of that range), and produces less acetaldehyde than most yeasts. The aromas are nice, and with an alcohol tolerance of 15%, this yeast is well-adapted to making sweet port-style wines. I like using it to make blueberry and cherry wines.

Lalvin D-47

If you'd like to make a dry white wine starting from apples or pears, this is an excellent choice. Its temperature range is narrow—only 58 to 68—but that makes it perfect for fermentations that proceed in the house during the winter when homes are usually maintained precisely in that range. The sediment formed by D-47 is compact, which makes racking easier.

Lalvin ICV-D254

With an alcohol tolerance of 18%, ICV-D254 will ferment any practical must to dryness. This yeast ferments quickly, so you'll want to keep the temperature under 80 degrees to avoid foaming. You might want to keep the temperature even lower to preserve volatile flavor components because ICV-D254 creates a very complex and fruity flavor profile that really enhances the fruit character of a wine. This would be a good choice for blueberry wine.

Wyeast 4632 Dry Mead Yeast

Meads, also known as honey wines, are enjoying a resurgence in popularity. Many yeasts will work to make mead, but this yeast in particular creates flavor notes that have resulted in many award-winning meads. The temperature range is 55 to 75, but you'll want to stay as close to 65 as you can to maximize flavor production. Wyeast 4632 has an alcohol tolerance of 18% and will result in a very dry mead.

Consolidated Ingredient List

The following ingredient list will allow you to make many successful gallons of wine. As your experience expands, you may wish to adopt different materials and techniques; but most home wine makers find that this list is more than sufficient for their needs. In compiling this list, I went to two well-known online retailers of winemaking supplies, and in both cases the total cost was under $40. You can save $7 by omitting the Wyeast #4632 from the list.

4 oz	Citric acid
4 oz	Malic acid
4 oz	Tartaric acid
2 oz	Liquid tannin
½ oz	Pectic enzyme liquid
2 oz	Yeast energizer
1 oz	Potassium sorbate
100	Campden tablets
4 oz	Powdered potassium metabisulfite
2 pkt	Red Star Pasteur Champagne yeast
2 pkt	Red Star Montrachet yeast
2 pkt	Lalvin D-47 yeast
2 pkt	Lalvin ICV-D254 yeast
1 pkt	Wyeast #4632 Dry Mead yeast

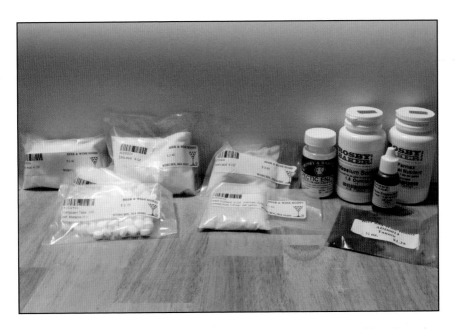

Important additives and adjuncts used for making wine. These are inexpensive and last a long time.

4

The Science of Wine

At its core, the theory of making wine (also beer and bread) is nothing more than the conversion of sugar into ethyl alcohol and carbon dioxide by the enzymes in yeast:

glucose → ethyl alcohol + carbon dioxide + energy

$$C_6H_{12}O_6 \rightarrow 2(CH_3CH_2OH) + 2(CO_2) + Energy$$

Using the foregoing formula based upon the molecular weights[5] of the compounds, 180 grams of glucose will be converted into 92 grams of ethyl alcohol and 88 grams of carbon dioxide. This means that the yield of alcohol, by weight, in a

---

5   The molecular weight of a compound is computed by adding the atomic weights from the periodic table of each constituent atom. The atomic weights of carbon, hydrogen and oxygen respectively are 12.0107, 1.00794 and 15.9994. So the molecular weight of glucose would be 6*12.0107 + 12*1.00794 + 6*15.9994. For ease of discussion I have rounded the results to the nearest gram.

perfect fermentation is 92/180 or 51%, and that nearly half of the weight of the sugar is lost in the form of carbon dioxide gas.

The density of glucose is 1.54 g/cm³, so the volume occupied by 180 grams of glucose is 180/1.54 or 116.88 cm³. The density of ethyl alcohol is .789 g/cm³, and the volume occupied by 92 grams of ethyl alcohol is 92/.789 or 116.6 cm³.

In other words, even though nearly half of the mass of sugar is lost in the form of carbon dioxide gas, the volume of the solution stays so nearly the same as to be indiscernible without resorting to very precise measurements.

Furthermore, the percentage of alcohol in beverages is not measured by mass, but rather by volume. This means that the volume occupied by alcohol in a completely fermented solution will be nearly identical to the volume of sugar that was in the solution. So if you know how much sugar is in a solution before fermentation starts, you know how much alcohol could be produced in a completed fermentation.

As we discussed a bit in the last chapter, what happens over the process of making wine is a lot more complex than a simple conversion of sugar into alcohol, so I'd like to expand on that some more.

Sugar

Unless you are using vitis vinifera grapes, all of your wine musts will contain less sugar than is needed to make a self-preserving wine. The sugar content of common fruits (other than wine grapes) is insufficient. In order for a wine to be self-preserving without need for pasteurization or the addition of preservatives, it needs an alcohol content of at least 9%. In practice, because you may add water between rackings in order to fill air space, you'll want enough sugar to yield an alcohol content of 10% or higher.

Measuring Sugar Levels

Many winemaking books and pamphlets are full of recipes that specify a certain fixed amount of sugar for a given fruit. Such recipes rely upon the false assumption that the sugar content of a given fruit is the same no matter how close to ripeness it was when harvested, how long it has been stored, or even the variety of the fruit in question.

The key to getting the sugar right is using a hydrometer. The hydrometer was discussed briefly in the previous chapter. As stated, it looks a lot like a

>> Using the weight method of determining specific gravity.

thermometer with a big bulb on the end. It measures the amount of dissolved solids in a solution by how far it sinks. There is a long stem and a scale, and the specific gravity is read where the liquid touches the glass. This is important because the surface tension of the liquid will give a false reading anywhere else, so be sure to read the value where the liquid is touching the glass.

I use a wine thief that doubles as a hydrometer jar. So I clean and sanitize the wine thief and hydrometer, and then give the hydrometer a spin as I put it into the liquid. Giving the hydrometer a spin is necessary because otherwise air bubbles could cling to it and give it false buoyancy that would give inaccurate readings.

Once you have your reading, you will need to correct it to compensate for the temperature of the must, because hydrometers are calibrated for 60 degrees. If the temperature is between 40 and 50 degrees, subtract 0.002 from the reading. If the temperature is between 50 and 55 degrees, subtract 0.001. If the temperature is between 65 and 75, add 0.001, and if the temperature is between 75 and 80, add 0.002. If the temperature is over 80, let it cool before measuring.

There is another method that I have never seen mentioned in books on wine-making, but I believe is superior even though it requires the use of math. The method is as follows:

Buy a jug of distilled water at the supermarket. Leave it at room temperature next to your primary fermenter so that it is at the same temperature (by doing this, you won't need to make temperature corrections later). Use a scale accurate to within 0.01g to weigh an empty and dry 10ml graduated cylinder.[6] Then fill the cylinder with 10 ml of the distilled water and record that weight. Finally, empty

---

6   I use the American Weigh AWS-100 scale. Complete with calibration weights it costs less than $20. Graduated cylinders are likewise ubiquitously available online for less than $10. Get a glass one rather than plastic. Large syringes are available from livestock stores and online for less than $3. You don't need the needle.

the cylinder, and using a large sterilized syringe, fill the graduated cylinder to the 10ml mark with wine must and record that weight. When measuring volume, put your eye at a level with the markings on the cylinder and fill until the lowest part of the liquid is perfectly aligned with the 10ml mark.

We now have three numbers. A is the weight of the empty cylinder, B is the weight of the cylinder filled with distilled water, and C is the weight of the cylinder filled with wine must. The equation for the specific gravity is: $SG = (C-A)/(B-A)$. For example, my graduated cylinder weighs 37.65g. When filled with distilled water it weighs 47.64g, and when filled with a light sugar syrup it weights 48.57g. $SG = (48.57-37.65)/(47.64-37.65)$ or 1.093. Because both the distilled water and the must were weighed at the same temperature, temperature corrections aren't needed. Even better, the amount of must used for testing was truly tiny—less than an ounce. Discard the sample in the sink after testing.

Adjusting Sugar Levels

As I mentioned earlier, in order for a wine to be self-preserving, it should have at least 9% alcohol. The following table gives you the potential alcohol based upon specific gravity, and how much sugar is present in a gallon of must to give you that much alcohol if it is completely fermented by the yeast.

Because water may be added to the wine at some rackings—thereby diluting the alcohol—you should also aim for a starting specific gravity that exceeds 1.080, corresponding to 10.6% alcohol. Also, even though a particular strain of yeast might have a theoretical alcohol tolerance exceeding 20%, such yeasts will not thrive in musts containing enough sugar to make that much alcohol. Higher levels of alcohol like that are achieved by fortification or by adding small amounts of sterile sugar syrup as existing sugars in the must are depleted. In order to avoid a fermentation failing due to excessive sugar levels, you should limit the initial specific gravity of your musts to no greater than 1.100, which corresponds to 13.6% alcohol. So, aim for a starting gravity between 1.080 and 1.100. In practice, I use 1.090 for almost all of my wines.

Almost all of your musts made from pressed or juiced fruits will contain insufficient levels of sugar to reach the minimum necessary alcohol content. Even though I am about to cover the math in more depth, the following shortcut equation will work fine:

Ounces of Sugar = (Desired S.G. - Measured S.G) x 360

Specific Gravity	Potential Alcohol	Ounces Sugar Per Gallon Must
1.000	0.0	0
1.004	0.6	1.4
1.008	1.1	2.8
1.012	1.7	4.3
1.016	2.2	5.7
1.019	2.6	6.8
1.023	3.2	8.2
1.027	3.7	9.6
1.031	4.3	11.0
1.035	4.8	12.4
1.039	5.4	13.9
1.043	5.9	15.3
1.047	6.5	16.7
1.050	6.9	17.8
1.054	7.4	19.2
1.058	8.0	20.6
1.062	8.6	22.0
1.066	9.1	23.5
1.070	9.7	24.9
1.074	10.2	26.3
1.078	10.8	27.7
1.081	11.2	28.8
1.085	11.7	30.2
1.089	12.3	31.6
1.093	12.8	33.1

Specific Gravity Table

If you decide to add honey rather than sugar, then multiply the amount of sugar needed by 1.3 to make up for the moisture content of honey. Ideally, you would use a scale for measuring sugar to be added; but if you don't have one, you can use measuring cups and allow for seven ounces of granulated sugar per cup.

For example, if you are making cyser from juiced apples and the measured S.G. of your must is 1.040 but you want a starting S.G. of 1.090, you first determine how much sugar is needed:

$$(1.090-1.040) \times 360 = 18 \text{ ounces}$$

Because you will be using honey instead of sugar, you'll multiply that by 1.3 to compensate for the moisture content of the honey: 18 x 1.3 = 23.4 ounces.

Sometimes, you might want to start with a high-quality bottled juice to make wine. You can tell how much sugar is in the juice just by reading the label, and it usually amounts to anywhere from 30g to 50g per 8 fl. oz. serving. Your first task in that case is to do unit conversion. Let me illustrate with an example.

I might have some 100% black cherry juice that I would like to turn into wine. It has 50g of sugar per eight-ounce glass. How many ounces of sugar does it have per gallon? A gallon is 128 ounces, so there are 16 eight-ounce glasses per gallon. So the total amount of sugar in a gallon of the juice is the amount in 16 eight-ounce glasses, or 16 x 50g = 800g. You convert grams to ounces by dividing by 28.35, so 800 g / 28.35 grams per ounce = 28.2 ounces of sugar per gallon. Looking at our table of specific gravities, we can see it already has plenty of sugar.

I may also have some organic concord grape juice that contains 40g of sugar per eight ounce glass. Doing the same math, (16 glasses x 40g)/28.35 = 22.6 ounces of sugar per gallon. That corresponds to only 8% alcohol, which is too low for a self-preserving wine. I want to bring it up to 12.3%, but to account for the increased volume from adding the sugar, I'll use the sugar quantity corresponding to 12.8% alcohol. So I need to add 33.1—22.6 = 10.5 ounces of sugar.

Sometimes using honey instead of cane sugar can give wine a really nice background flavor. When using honey as a substitute for sugar, just multiply the number of ounces needed by 1.3 to compensate for the honey's water content. Many cookbooks advocate oiling the containers used for handling honey. Do not do this, as you'll end up with a persistent oily layer in your wine. Instead, heat the honey by placing the jar and any handling tools in simmering water. That will make it easier to use without adding oil to your must.

To give you some idea of how much sugar would need to be added to the juices of various fruits, I have included a table listing some fruits and the range of specific gravities I obtained when testing different varieties. Keep in mind that this is just a guideline. Don't substitute use of this table for testing the specific gravity yourself because the particular fruits you use will be of different varieties, grown in other places, and harvested at different times.

Fruit	S.G. Range	Fruit	S.G. Range
Apples	1.040—1.060	Blackberries	1.020—1.035
Blueberries	1.045—1.055	Currants	1.042—1.060
Cherries (sweet)	1.045—1.075	Cherries (tart)	1.040—1.070
Cranberries	1.015—1.020	Grapefruit	1.028—1.041
Lemon	1.025—1.050	Peach	1.030—1.040
Pear	1.040—1.045	Pineapple	1.045—1.060
Plum	1.045—1.055	Black Raspberry	1.030—1.050
Strawberry	1.020—1.040	Watermelon	1.030—1.040

Specific Gravity Ranges of Common Fruits

Tannins

Tannins are complex polyphenols[7] produced by plants. In foods, they are bitter and astringent, and it is theorized that they serve to deter herbivores, though it is likely that they serve other purposes as well. Chemically, they can be divided into several categories, but they all have in common the characteristic that they are able to bind to proteins and precipitate them out of solutions.

Tannins are more present in the skins, seeds, and woody portions of plants. Hence, when red wine is made by fermenting the seeds and skins of the grapes, tannin is dissolved into the wine. The solubility of tannin is affected by the pH of the solution. Tannins are more soluble in neutral solutions than in acidic ones, and they are more soluble in alcohol than in water. So when a fermentation first starts, very little color and tannin is extracted but once some alcohol is produced, the extraction proceeds more rapidly. In addition, tannins are imparted to wines through aging processes that utilize oak barrels or the addition of oak cubes to the fermenter.

The ability of tannins to precipitate proteins has important implications for the aging of wines and beers. Precipitation refers to the dissolved tannins combining with dissolved proteins to form a compound that can't be dissolved. This compound, once formed, slowly sinks to the bottom of the vessel. When this

---

7   Phenol is a benzene ring compound with an -OH group on one of the carbons, making it an alcohol. Polyphenols are compounds composed of multiple phenol groups bonded together. The molecular weights of tannins range from 500 to more than 3000.

happens, the astringent or bitter flavors imparted by the tannin are lessened and the haziness imparted by the protein is diminished.

Tannins are also chelators. That is, they combine with the ions (positively charged atoms) of metals in order to make them non-reactive. A major effect of tannins is that they combine with iron in such a way as to make it biologically useless to living things. Pathogenic bacteria love iron. They love iron so much that they invade the human body to get it.[8] One of the reasons why red wines keep so much better than others is because the tannins have tied up the iron, making the environment unattractive for pathogenic bacteria. Tannins also chelate magnesium, copper, and other metals, but do so without making the metals unavailable. This alters the taste by altering the nature of the compounds.

Ingredients other than grapes can have more or less tannin content, and that content will vary based upon the amount of time the whole fruit is left in the primary fermenter as well.

Unfortunately, because tannins encompass such a vast array of compounds, assessing the tannin content of a must is a devilishly complex exercise in experimental chemistry. If you are curious, please see *New Tannin Assay for Winemakers* by Moris L. Silber and John K. Fellman for the most accurate method using protein dye markers or the older (and more controversial) precipitation technique published by Hagerman and Butler in the *Journal of Agricultural Food Chemistry* in 1978.

Some fruits already have so much tannin that they should be diluted in order to make a drinkable wine, whereas others will require the addition of tannin to help pull proteins out of solution. I have included a table of common fruits that shows how much relative tannin they have, divided into low (less than 3 grams per liter), medium (3–4 grams per liter) and high (more than 4 grams per liter). If a juice is in the "low" category, add ⅜ teaspoon of grape tannin per gallon. If a juice is in the "medium" category, add ¼ teaspoon per gallon. If it is in the "high" category, you will likely need to dilute the juice with water or a juice with lower tannin content to avoid making a wine that is too astringent to be enjoyable. If you have to dilute the juice anyway because of its acidity (later in this chapter), consider the diluted juice to be one category lower for purposes of tannin content.

---

8   Ewald, Paul (2002), *Plague Time: The New Germ Theory of Disease*

Low Tannin	Medium Tannin	High Tannin
Apples, Bananas, Cranberries, Lemons	Blueberries, Blackberries, Cherries (sweet), Currants, Gooseberries, Grapes, Grapefruit, Passionfruit, Plums, Raspberries, Strawberries	Apricots, Blueberries, Cherries (sour), Guava, Kiwi, Mango, Oranges, Peaches, Papaya, Pears

Relative Tannin Content of Common Wine Ingredients

Your fruits will certainly differ to some degree from those I used for testing and my testing method used my own home lab rather than a professional lab, so I recommend that you mix up your must and add half the tannins specified, and then take a clean spoon and actually taste the must. If it isn't giving you any "pucker effect" go ahead and add the rest of the tannin specified. A must that starts out tasty will likely turn into a tasty wine!

If you have a finished wine that for some reason has excessive tannin, keep in mind that some wines are at their best after being stored for several years, during which time the tannins slowly polymerize, combine with proteins or otherwise become less astringent. If that doesn't work, or you need to use a wine early, you can precipitate out the tannins using a combination of gelatin and kieselsol.

To use gelatin, use your scale to measure out one gram of fining gelatin (from a winemaking store), and mix that with two tablespoons of cold water in a clean coffee cup. Separately, put seven tablespoons of water in a glass measuring cup, and heat on high in the microwave for one minute. Add the hot water to the dissolved gelatin in the coffee cup, mixing thoroughly. Allow this to cool down to a temperature of 80 degrees, and then gently stir in two tablespoons per gallon of wine or the whole amount for five gallons. Leave it for two to five days before adding the kieselsol.

Whenever you use gelatin, it will impart some haze to the wine. This can be removed with kieselsol, a soluble silica gel. Soluble silica gel has an ionic charge that will attract uncombined gelatin and gelatin-tannin complexes. This will precipitate quickly. Use one ml per gallon of wine. Stir it in gently. Wait at least five days but not more than ten before racking the wine to leave the precipitated tannins behind. (Racking is explained in the next chapter.)

Acids in Wine

The acidity of wines is important because the organic acids help establish an environment favorable to yeast. They also combine over time with alcohols to enhance flavor and smell, and they assist sulfite in sanitizing the must. Most importantly, they convey a taste of their own that balances the wine.

Depending upon the fruit you use, your wine must will already contain a combination of organic acids. Every fruit has some amount of citric acid, as citric acid

Fruit	Average Acidity in grams/liter	Primary Acids
Apple	6.5	Malic, citric, lactic
Banana	3	Citric, malic, tartaric
Blackberry	13	Malic, citric, isocitric
Blueberry	13	Citric, malic
Cantaloupe	2.5	Citric, malic
Cherry (sweet)	11	Malic, citric, isocitric
Cranberry	30	Citric, malic, quinic
Grapefruit	20	Citric
Grape	6	Tartaric, malic
Guava	12	Citric, malic, lactic
Lemon	40	Citric
Mango	3	Citric, tartaric
Orange	15	Citric, malic
Papaya	0.5	Citric, malic, ketoglutaric
Passion Fruit	25	Citric, malic
Peach	8	Malic, citric
Pear	4	Malic, citric
Pineapple	10	Citric, malic
Plum	6	Malic, quinic
Raspberries	14	Malic, citric, isocitric
Strawberries	10	Citric, malic
Watermelon	2	Citric, malic

Acidity of Common Wine Ingredients

is crucial to metabolism, but often a different acid is predominant and the combination of acids is unique for every fruit. Each fruit also has a different overall level of acidity. Some fruits are so acidic (> 9 grams per liter) they cannot be used exclusively to make a wine must, and their juice must be diluted with either water or the juice of a less acidic fruit. The following table lists common fruits, their acidity as tested by titration and the primary organic acids in each fruit in decreasing order of relative quantity.

Measuring Acidity

Thankfully, unlike tannins, which are hard to measure, the overall acid content of wine musts is easy to determine. Wine musts contain a variety of acids, but it isn't possible for a home winemaker to separate these out and measure them independently. Because each of the primary organic acids has a different molecular weight (150.9 for tartaric, 134.1 for malic and 192.1 for citric), but a mole of each is neutralized by two moles of sodium hydroxide, what winemakers have standardized upon is interpreting the results of the tests as though the acid being neutralized were tartaric. Likewise, wine makers don't usually like to think in terms of moles, so the results are converted via a multiplier into the more familiar "parts per thousand." So acid measurements of wine must are provided in terms of TA (titrateable acidity) as tartaric in PPT (parts per thousand). This is the same thing as grams per liter, abbreviated as g/L.

The method of measuring the acid content is called titration, and it takes advantage of the fact that acids and bases neutralize each other. You might have observed this phenomenon as a kid by mixing baking soda with vinegar. The combination generated carbon dioxide gas initially, but after a while settled down and did nothing once either component was fully neutralized. We won't be using baking soda because we don't want to generate gas. Instead, we'll use a standardized solution of sodium hydroxide—otherwise known as lye.

The widely available acid test kits have a problem. That is, they rely upon the color change of an indicator (phenolphthalein) which turns pink when enough sodium hydroxide has been added. But if you are dealing with a pink, blue, or purple sample, ascertaining when it has changed color is really difficult. I recommend using an inexpensive pH meter[9] instead.

---

9   I use the Milwaukee pH600. It costs about $20 from various vendors.

By using either an indicator that changes color when the solution has been neutralized or a pH meter, you can tell when enough base has been added to neutralize the acid. Because you know the concentration of the base you are using, the amount of acid in your test sample can be easily calculated. The calculation is as follows:

(Normality of Base) X (Volume of Base) = (Normality of Unknown) X (Volume of Unknown)

Because the calculations are just the arithmetic of converting molarity to grams per liter, I have designed the procedure below to take that into account, and use just a one-time multiplication.

Supplies

150 ml beaker
1 glass stirring rod
110 ml syringe
1 cup distilled water
1 container of 0.1N sodium hydroxide solution

Procedure

Wear safety glasses.

Use the clean syringe to measure 5 ml of wine must and transfer it into the beaker.

Clean the syringe and then rinse with the distilled water.

Fill the syringe to the 10ml mark with sodium hydroxide solution

Add the sodium hydroxide to the beaker 0.1 ml at a time. After each addition, stir the contents of the beaker and test the pH with the meter.

Repeat the previous step until the pH meter reads 8.3 or higher. Then stop.

Make note of the reading on the syringe.

The TA (tartaric) in PPT (or g/L) of your must is equal to 1.5 x (10—reading on syringe).

Clean, rinse, dry, and store your equipment.

Adjusting Acidity

Acids affect flavors and indirectly create new flavors in a maturing wine. When making wine, the acidity of a must needs to be adjusted so that it is high enough, but not so high as to make an unpleasant flavor. Though your sense of taste is the final arbiter, there are some ranges of acidity that have been established by wine makers over time that can serve as a general guideline:

Dry White Wine: 7.0-9.0g/L
Sweet White Wine: 8.0-10.0g/L
Dry Red Wine: 6.0-8.0g/L
Sweet Red Wine: 7.0-9.0g/L
Dry Fruit Wines and Meads: 5.0-6.5g/L
Sweet Fruit Wines and Meads: 6.5-9.0g/L
Sherries: 5.0-6.5g/L

Many country wines are blended. For example, you might make a blueberry wine that contains a fair amount of red grape concentrate. So consider the full nature and character of your wine in assessing which category of acidity is appropriate. In the case of a dry blueberry wine containing red grape concentrate, I'd be aiming for about 7.0g/L.

If you find your wine is too acidic, no more than 2g/L too much, you can reduce the acidity by adding potassium carbonate. Potassium carbonate has a molecular mass of 138.2, and tartaric acid has a molecular mass of 150.9. There are 3.79 liters in a gallon, and potassium carbonate removes one molecule of acid for every molecule of potassium carbonate added, so for every PPT reduction in acidity required, add 3.5 grams of potassium carbonate per gallon.

For example, if I have five and a half gallons of wine must as described above, it has an acidity of 8.2g/L and I want an acidity of 7.0g/L, the amount of potassium carbonate I would need to add is:

(5.5 gallons) x 3.5 grams (8.2g/L−7.0g/L) = 23.1 grams. Measure it with your scale for best accuracy.

For reductions greater than 2.0g/L, I do not recommend adding potassium carbonate as it can impart undesired salty tastes. Instead, I recommend blending. You can blend with water or other juices with lower acidity. In general, you don't want to blend with too much water as that will reduce the flavor and increase the amount of sugar you'll need to add. Keep in mind that whatever fruit juices you use for blending shouldn't overpower the primary ingredient. This will require a bit of algebra.

Pretend I want to make blueberry wine. To that end, I have juiced some blueberries, and tested the acidity of the must at an excessively sour 11g/L. I want 7g/L. I am making 5.5 gallons of must.

There are 3.79 liters in a gallon. If my desired acidity is 7g/L, then the total amount of acid in 5.5 gallons of must will be 7g/L X 3.79L/gallon X 5.5 gallons = 145.9 grams. My blueberry must contains 11g/L of acidity, which works out to 11g/L X 3.79L/gallon = 41.7 grams per gallon.

If I wanted to dilute the juice with water alone, it would be easy to determine how much blueberry must I could use by dividing the total amount of desired acid in 5.5 gallons of must (145.9 grams) by the number of grams of acid in a gallon of my blueberry must (41.7 grams). So 145.9 grams / 41.7 grams per gallon = 3.5 gallons. So to make 5.5 gallons of must with the proper level of acidity, I would use 3.5 gallons of blueberry must and make up the remaining two gallons with water. Because blueberries are very strongly flavored, this would likely work fine as long as we added tannin and sugar as needed.

Of course, we wouldn't have to use water. We could use watermelon juice instead! If we have one gallon of watermelon juice has an acidity of 3g/L, how much blueberry juice and water would we need to use?

The total acidity available from the watermelon juice is 3.79L/gallon X 3g/L X 1 gallon = 11.4 grams. The must requires a total of 145.9 grams, so the amount of acidity remaining is 145.9 grams–11.4 grams = 134.5 grams. If we divide that by the number of grams of acid per gallon of blueberry juice (41.7 grams) we get 134.5 grams / 41.7 grams per gallon = 3.22 gallons. That's close enough to three gallons plus a quart, so now our recipe is 3.25 gallons of blueberry juice, one gallon of watermelon juice, and the remaining (5.5 gallons–3.25 gallons–1 gallon) 1.25 gallons made up with water. As you can see, the math for blending to get the right acid levels isn't very difficult.

Usually, however, excessive acid is not the problem. The problem is more likely to be insufficient acid. This is especially the case with low or medium acid fruits that are fully ripened, and with fruits whose quantities need to be kept low due to high tannins such as cherries.

If I were making a cherry wine, because cherry is high in tannin, I would likely use half cherry juice and half red or white grape juice in my must. Because the result would be a red wine, I'd want the acidity to be at around 7.0g/L. In all likelihood, though, when I measured, I'd find the acidity closer to 5.5g/L.

To increase the acidity, you add acid directly to the must and stir it in. Winemaking shops make citric, tartaric and malic acids available, as well as an acid blend composed of equal parts of all three. The only place I can see acid blend being used is in meads (honey wine) that have no fruit component. Otherwise, what I recommend is the use of acids based upon the nature of the fruit.

Earlier in this chapter is a table that lists, in order of influence on taste, the primary organic acids present in a variety of fruits. For some fruits, the primary acid is malic, for others it is citric or tartaric. When correcting the acidity of a must whose primary character is that of a particular fruit, you should use the two most important acids for that fruit in a 2:1 ratio.

For example, if I am making an apple wine, the primary acids are first malic and then citric acid. When I add acidity to the must, I will add a blend of acids composed of two parts malic acid and one part citric acid.

Determining how much acid to add is straightforward. If I want my must to have 6.5g/L acidity and it only has 5.0g/L of acidity, then I need to add 6.5g/L−5.0g/L = 1.5g/L of acid. Converting that to gallons simply requires multiplying the result by the number of liters in a gallon, which is 3.79. So to increase the acidity of 5.5 gallons of must from 5.0g/L to 6.5g/L I would need to add 1.5g/L X 3.79 L/gallon X 5.5 gallons = 31.3 grams of acid. In even numbers, then, I would add 20 grams of malic acid and 10 grams of citric acid.

There is a school of thought that citric acid should never be used in wine musts. The reason is because citric acid can promote acetification (i.e. the process of turning wine to vinegar) or can contribute to the development of diacetyl (buttery) flavors. Both statements are true. However, if you are scrupulous in your sanitation, acetification is not likely to happen and some wines could benefit from any diacetyl developed. That having been said, if you are concerned about this, you can substitute tartaric acid for citric acid, and by doing this you will increase the grape flavors in your wine.

Pectins

Pectins are long chains of carbohydrates composed of various sugars that form the cell walls of the fruits used to make wine. Pectins are responsible for turning the juices of some fruits into jelly. European grapes contain enough pectic enzyme—an enzyme that destroys pectin—to destroy that pectin so you end up with a clear fluid wine rather than a semi-solid gelatinous mass. Other fruits don't usually have enough of this enzyme naturally, which makes them excellent for making jelly but suboptimal for wine.

Pectic enzyme purchased from the winemaking store is used in small amounts to supplement the natural pectic enzymes in the must. Over time, this degrades the pectin and thereby either makes its sugars available for fermentation or precipitates the leavings into the bottom of the fermenter so they are left behind at the next racking. Therefore, pectic enzyme helps to produce clear wines.

You may recall that one reason most fruits and vegetables are blanched before freezing or dehydrating is that the high temperature of blanching inactivates the enzymes that cause the produce to degrade over time. The same will occur with pectic enzyme, so pectic enzyme should only be added to a must with a temperature under 80 degrees and the must cannot be reheated thereafter.

When making wines with no added sulfites or when making wines in which honey is the primary ingredient, it is common to heat the must in order to assure its sterility. Anytime heated ingredients are added to the must, the temperature of the must should be allowed to drop adequately before pectic enzyme is added. The container of pectic enzyme has instructions printed on the label for how much to add to your must, but this is a general direction. Some fruits require the standard amount, but some require double. The following table will let you see at a glance.

Enzyme According to Directions	Double Pectic Enzyme
Blackberries, Blueberries, Cherries, Nectarines, Peaches, Plums, Raspberries, Watermelon	Apples, Pears, Strawberries

Pectic Enzyme Requirements of Various Fruits

Yeast Nutrient, Yeast Energizer, Thiamine, and Lipid Supply

During the reproductive phase of yeast in the must, the sheer volume of yeast that is created from a tiny packet is impressive. There will literally be millions of yeast cells per milliliter of must. All of this cellular budding and division requires core building blocks for protein and the other parts of a yeast cell. As with many important factors, though these are usually present in European grapes to a sufficient degree, they are lacking in practically all other primary ingredients for winemaking.

A wine can be made successfully in some cases without the addition of nutritional building blocks for the yeast, but adding those building blocks will go a long way toward stacking the deck in favor of a successful outcome.

You will see wine supply stores selling many supplements for yeast with names such as yeast nutrient and yeast energizer. There is no universal standard,

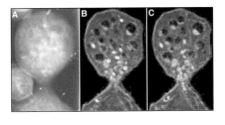

» Yeast cell division requires proper nutrition.

and so the precise ingredients will vary with the supplier. In general, they will contain purified sources of nitrogen and phosphorus at a minimum, though many will also contain a variety of B-vitamins. Yeast nutrient usually contains only food grade ammonium phosphate, whereas yeast energizer will contain this along with magnesium sulfate, killed yeast, and the entire vitamin B complex; of which thiamine (vitamin B_1) is the most important. Sometimes you may see urea as an ingredient. If you do, don't worry. This is purified food-grade urea that supplies nitrogen for building proteins and it is perfectly safe.

I would recommend using yeast energizer in preference to yeast nutrient. But if you use yeast nutrient instead, at least add a 100% RDA thiamine tablet and a pinch of Epsom salt in addition for each gallon of must.

The cell walls of yeast also require lipids (fats), and such fats are in short supply in some wine musts—especially meads made predominantly from honey with little or no fruit. In such cases, you can use yeast hulls as an additive or a specialized additive that contains essential fatty acids such as FermaidK or Ghostex.

Sulfite

Some people who get headaches from drinking wine believe themselves to be sensitive to sulfites. Usually, however, they get headaches from red wine but not from white wine, both of which contain sulfites. So sulfites are not the issue.[10] This headache is called Red Wine Headache, and experts disagree widely on its true cause. Less than 1% of people are truly sensitive to sulfites which are found ubiquitously in lunch meats, dried fruits, and even white grape juice from the supermarket. Obviously, if you are truly sensitive to sulfites you should avoid them at all costs.

Sulfite is used so pervasively in winemaking and considered so essential that its use is even permitted in wines labeled as USDA Organic. Though it is possible to make wines without the use of sulfites and I have successfully done so, the odds

---

10 K. MacNeil, (2001) *The Wine Bible*

of success for a beginner are greatly enhanced by using sulfites, especially if you are using fresh or frozen fruits in the must.

Sulfite is used in winemaking as a sanitizer to kill or inhibit wild yeasts and bacteria so you end up making wine instead of vinegar. It is also used to help clear wines during racking to arrest fermentation and to help prevent oxidation and consequent degradation of flavor.

Sulfite comes in many forms, but for our purposes two forms are important. The first is potassium metabisulfite in the form of Campden tablets. Campden tablets are sized with the idea in mind of accurate dosing of wine and musts to purify must prior to initiating fermentation and help clear and preserve the wine later. To use Campden tablets, do not just plunk them into the wine or must. Instead, use a cleaned and sanitized wine thief to remove four to eight ounces of must or wine, and put it into a sanitized glass. Thoroughly crush the requisite number of tablets, and add the powder to the must or wine. Stir to dissolve. Once the tablets are dissolved, add the must or wine back to the original container. For the initial sanitizing of a must, use two tablets per gallon of must. For protecting wine from spoilage and oxidation, add one tablet per gallon before racking.

The second is powdered potassium metabisulfite. In powdered form it is used to make sterilizing solutions for sterilizing equipment.

Make a gallon jug (a clean empty plastic water jug is fine) of sanitizing solution. To make the sanitizing solution, dissolve a measuring teaspoon of potassium metabisulfite powder in a gallon of water. You can use this solution repeatedly, and pour it back in the bottle after each use to rinse a fermenter or a racking tube until it loses its potency or becomes obviously dirty. If you keep the container tightly sealed when not in use, it will stay effective for a very long time. You can tell if it is potent by sniffing the solution. If the scent just barely tickles your nose, it needs to be replaced.

There are other sanitizers available and when you have become more experienced and confident, you can branch out and start experimenting. But sulfites are the easiest to use not only for the beginner, but also for the most prestigious of professional wineries.

Testing Sulfite Levels

It is very rare that you would need to test the sulfite levels in wine. Simply following the directions in this book will assure adequate but not excessive levels of sulfite for most purposes. However, there are instances where you'll want to know how much sulfite is in the wine. For example, if you plan to follow your secondary fermentation with a malolactic fermentation in order to reduce perceived acidity, most malolactic cultures will be inhibited by sulfite levels greater than 20 ppm. So if you have been dosing regularly with sulfite between rackings, when you rack into a container to initiate malolactic fermentation, you should test the sulfite levels in your must, and reduce them if they are too high.

You can purchase sulfite test kits from wine equipment suppliers. These test kits use what is called the "Ripper" method and they work quite well with wines that are not strongly colored. With strongly colored wines, they give a reading that is too high because the compounds that impart color to the wine combine with some of the test ingredients making them inert. You can "guesstimate" the error by subtracting 10ppm from the results of the test, or you can do a more elaborate test on your own. I've detailed that test in the advanced techniques chapter.

To reduce sulfite levels, stir 3% hydrogen peroxide solution from the pharmacy into your wine and wait an hour. The amount you need is equal to 1 ml for every ppm reduction per gallon of wine. So if I have 5 gallons of must with a sulfite level of 33 ppm and I want to reduce the sulfite level to 15 ppm before adding a malolactic culture, I need a reduction of 33ppm–15 ppm or 18 ppm. The amount of 3% hydrogen peroxide solution to add is calculated like this:

(Gallons of wine) X (ppm reduction desired) = ml of 3% hydrogen peroxide solution to add

So 5 gallons X 18ppm = 90ml of 3% hydrogen peroxide solution.

Yeast

Yeast is the star of the show. Wild yeast naturally colonizes the surfaces of fruits, so sometimes crushed fruit, left to its own devices and protected from other organisms, will ferment all by itself. In fact, this is the case in certain famous wine regions where the wild yeasts inhabiting the area have co-evolved with the wine grapes. Though most wine yeasts are of the species *Saccharomyces cerevisiae*, there are hundreds if not thousands of variations of this species, some with dramatically different properties. The genome of wine yeast has over twelve million base pairs, making for substantial possibilities for variation.

In practice, wine makers do not rely on wild yeasts because the unpredictability can often result in serious failures or faults in the finished product. Instead, wine makers usually purify the musts of wild yeasts and bacteria by adding sulfite.

Once the sulfite has been added, the must is stirred thoroughly and then allowed to sit for a day before a cultured wine yeast of known character is added.

Adding yeast to the must is known as pitching the yeast, though in reality little real pitching occurs because one- and five-gallon batches are relatively small. In batches of this size, the packet of yeast is just sprinkled as evenly as possible on top of the must in the primary fermenter. Do not stir. If you stir, it will take the yeast far longer to multiply enough to become active. You want the yeast to become active as quickly as possible because it is added after the sulfite has dissipated, so long lag times expose your must to a risk of spoilage by delaying the onset of production of alcohol.

Because yeast needs oxygen in its initial replication stage before fermentation begins, you should aerate the must by stirring it vigorously before pitching the yeast. Some wine makers put a sanitized fish tank aerator connected to an air pump into the must for an hour or so before pitching the yeast, but I have found a good vigorous stirring (carefully so as to avoid sticky messes) to be sufficient.

Yeast comes powdered in packets, in liquid in vials, and in many other forms. As you become a more advanced wine maker, you might decide to use liquid yeasts. The liquid yeasts require amplification, which I have covered in the chapters on beer. But your initial use should be of powdered dry yeast in individual foil packets. These are very well-characterized and foolproof. Just open the packet and sprinkle on top of the must—and it works. Don't be fooled by the simplicity of use or the fact these yeasts are inexpensive. Dry wine yeasts are a very high quality product and I have used them successfully for years. If you skimp and use bread or beer yeast to make your wine, don't complain if your wine tastes like bread or is syrupy-sweet because the alcohol tolerance was too low.

5

Ingredients and Techniques

Theoretically, you could make "wine" or at least a liquid containing alcohol, from just sugar, water, yeast and some nutrients. But the whole point of wine is to preserve the nutritional content of the starting fruits or vegetables, so we'll look at it from that point of view. Any fruit or vegetable can be used to make wine. Other than wine grapes, all fruits will require some amount of supplemental sugar. The juice of some fruits will require considerable dilution due to their high degree of acidity or astringency, and some will produce wines so tasty you'll wonder why you can't find them commercially. Others, like asparagus, will be downright unpalatable in some cases and suitable only for making marinades.

Fruits

Wine grapes are the perfect fruit for making wine. All you need to do is crush them and they make the perfect amount of juice with the perfect levels of acidity and sugar. Every other fruit is imperfect in some way. While fruits other than grapes are imperfect, they can be made perfect through proper adjuncts and technique.

With air transportation for produce so prevalent, there are more fruits available in our local markets than I could ever list, and quite a few I haven't tried because they are so expensive, such as starfruit and guava. In general, the higher the quality of the fruit, the higher the potential quality of your finished product. You will never make great wine from bad fruit—no amount of technique will improve its quality. But if you start with the highest quality fruit, there is at least the potential for creating great wine through solid technique.

You can use fruit in nearly any form to make wine: fresh fruits, dehydrated fruits, canned fruits, and frozen fruits. Fresh fruits and frozen fruits give the best results, and in many cases frozen fruits are superior to fresh because the process of freezing breaks down the cell walls to release more juice and flavor. Canned fruits often have a distinctly "cooked" taste that can detract from a wine, making it taste flat. They are best used for no more than half of the fruit in a wine. Wine-making shops sell specially canned fruits that come out better in wines than the canned fruit at the supermarket, but even these should constitute no more than half of the fruit by weight.

Dehydrated fruits retain their sugar, but have been subjected to oxidation and the loss of some of their more volatile flavor components. Usually, they are used in the form of raisins for purposes of adding some grape components to a wine so that it has a more vinous quality; dehydrated fruits in general, such as prunes and apples, are good for adding sherry-like taste qualities. Dehydrated banana is good for adding body to a wine such as watermelon wine that would otherwise be very thin. Very often, dehydrated fruits are sulfited to preserve their color. This is not a problem when they are added to a primary fermenter. In general, one cup of minced dried fruit will impart three ounces of sugar to the must, but this rule of thumb is no substitute for measuring with a hydrometer. Do keep in mind that making wine out of a dried fruit can concentrate the effects of that fruit, as I found to my chagrin with some prune wine I made.

Fruits, you will discover, are pretty expensive in the quantities you'd use for making wine. For example, you'll need twenty pounds of blueberries to make five gallons of blueberry wine. If you buy frozen organic blueberries at the super-

❯❯ Always use unwaxed fruit. Waxed fruits will create a mess rather than wine.

market for $3.69/lb, that means $73.80 just in fruit. Since you get twenty five bottles of wine from five gallons, that works out to just under $3/bottle, which is still a decent deal. Even so, it quickly becomes clear that your best bet is to either grow fruit yourself, go to a pick-it-your-self place or buy it in bulk from a farm stand. I pick the blueberries for my wine at Mrs. Smith's Blueberries nearby, and it's a lot more affordable. (You can also make wine in one-gallon batches so your initial outlay isn't so much. This is a good idea when experimenting!)

Fresh fruits for country wines are primarily processed using only one technique. In this technique, the fruit is placed in a clean nylon straining bag in the bottom of the primary fermenter, crushed with cleaned/sanitized hands, with the difference in volume being made up by adding water. The water helps to extract the dissolved sugars and flavor compounds, and as fermentation begins, the alcohol created helps to extract the color. This technique is best suited to softer fruits that are easily crushed by hand, though it is used for practically all fruits for the overwhelming preponderance of country wines.[11]

As an alternative, especially for harder fruits such as apples, I recommend using a high quality juice machine such as the Juiceman™ or Champion™. With these machines, the expressed juice goes into one container and the pulp goes into another. For darker fruits from which you want to extract color, such as cherries or blueberries, scoop the pulp from the pulp container into a nylon straining bag that you put in the bottom of the primary fermenter. (Note: Exclude the pits from stone fruits as they contain a cyanogenic glycoside that is poisonous.)

---

11 If you are using purchased fruits, please make sure they are un-waxed. The wax that purveyors use to make fruits look pretty will turn your intended wine into a useless mess.

Juices

I have made very good wines from high-quality bottled juices. For example, two quarts of apple plus two quarts of black cherry with the sugar and acid levels adjusted and a hint of vanilla added will make a gallon of really great wine.

Bottled juice that hasn't been treated with an additive that suppresses fermentation (such as potassium sorbate) can also be used to make wine. Keep in mind that something like the generic apple juice you can buy cheaply by the gallon is hardly more than sugar water and doesn't make very good wine. But there is a big difference between brands, and sometimes you can make a really excellent wine out of a blended Juicy Juice™.

Bottled juices and juice blends from the natural food section of the grocery store are often 100% juice from the described fruit. These have been specifically formulated to retain the distinctive flavor of the fruit, and can be easily used as an addition to wines. You might want to be sparing in their use though, as they often cost as much as $10/quart.

Grape juice concentrates can help add "vinous" quality to a country wine, making its mouth-feel resemble that of traditional wines. These are special concentrates purchased from winemaking stores that have had the water removed under vacuum, and have been preserved with sulfites rather than through heat; therefore, they preserve a distinctive grape character. At roughly $16/quart (they make a gallon of must when water is added) they are expensive, but they make good additions as part of a must. They come in white and red varieties.

Vegetables

Vegetables are used for wine either by boiling them in water and including the water in the must, or by juicing the vegetables with a juice machine. Many vegetables, no matter how they are handled, will impart a haze to wines, but this effect is more pronounced when using boiled vegetables. This is because boiling tends to set the pectins while denaturing the natural enzymes in the vegetable that would otherwise break down the pectins over time. There's no reason why you couldn't try bottled vegetable juices so long as they haven't been treated with a fermentation inhibitor, but the results can be pretty iffy when using brands that include added salt. Salt is added to vegetable juice to balance natural sugars for a tasty beverage. But when you use salted vegetable juice in wine, the sugar is converted to alcohol during fermentation but the salt remains. The results can be good for making marinade but decidedly not good for drinking. On the flip side, there's nothing wrong with having a variety of self-preserved marinades ready and waiting!

Speaking of marinades, both wines and vinegars are commonly used for this purpose, and both are self-preserving. You can make very good marinades by fermenting mixtures that include onions, herbs, celery, parsley, and similar ingredients. With their high alcohol content, they will keep for decades.

If you aren't making marinades but you are instead looking to make drinkable wines, both carrots and tomatoes can be excellent candidates for a wine. Carrots also blend quite well with apple. But don't let the fact that I've never made okra wine deter you if you want to give it a try.

Herbs and Spices

Though spices are not added to wines very much today, in past times spices were quite expensive so heavily spiced wines were an indicator of wealth and status. Unlike the traditional wine makers of France, as a home wine maker you don't have to contend with the traditional rules for making wine. One bonus is that you can add anything you'd like. You can add mulling spices to an apple wine, a hint of vanilla and cinnamon to a blueberry wine, and just a touch of rosemary to a carrot wine. The only rule is to make something that you and your friends will enjoy drinking, so if spices can enhance a wine to your tastes, then there's no reason not to use them. However, just as with food, it can be easy to over-do spice. Better too little than too much.

When adding spices, use whole rather than powdered ingredients. For one thing, powdered spices tend to have lost some of their volatile flavor components and will give inferior results. For another, they often form a haze in wines that is harmless but unsightly.

The technique for use is straightforward. Put the chosen spices in a spice bag, and lightly boil the bag in a quart of sugar water for ten to fifteen minutes, then discard the bag. Allow the spiced sugar water to cool to room temperature before adding to the must.

Spice	Goes best with	Amount to use
Peppercorns	Used to add warmth to most wines	5–10 whole peppercorns per gallon
Cassia Buds	Apple, blueberry, cherry, and most fruit wines	10–30 buds per gallon
Cinnamon	Fruit wines	1 stick per gallon
Cloves	Fruit, vegetable, and grain wines	3–6 cloves per gallon
Allspice	Fruit, vegetable, and grain wines	4–10 berries per gallon
Nutmeg	Fruit and vegetable wines	No more than ½ meg per gallon
Ginger	Best in lightly flavored wines such as apple and carrot	2–8 ounces, grated
Star Anise	Fruit wines	1 star per gallon
Vanilla	Fruit wines	1 vanilla bean or less per gallon

Sources of Sugar

Because yeast contains enzymes that turn many forms of sugar into a sort more easily used, any common source of sugar will have the same result in terms of alcohol production. You can use granulated cane sugar, dextrose, glucose, fructose, honey, molasses, brown sugar, maple syrup, high-fructose corn syrup, dried fruits, concentrated fruit juices, and more.

Though the source doesn't matter in terms of creating alcohol, it can make a big difference in terms of taste—for example, many of the chemical compounds that make honey or brown sugar have a distinctive taste and aroma which will be preserved in wines that include them. For this reason, I would recommend against using brown sugar or molasses.

Glucose, dextrose, fructose, and sucrose (cane sugar) are all treated identically by yeast. If the sugar isn't in a form the yeast can use, the yeast employs an enzyme called invertase to change it into a usable form. Nothing is gained by using the more expensive fructose from a health food store over an inexpensive bag of granulated sugar from the grocery store. None of these contribute flavor to the wine, and simply serve as a source of sweetness or alcohol. They are a good choice for wines in which you want the tastes and aromas of the primary fruit to dominate their character.

Bottled juices and juice concentrates can also be used as a source of sugar, especially given that sugar is their primary solid constituent.

Containing a wide array of minerals, amino acids, and vitamins, honey is a tasty addition to many wines. A number of cultural traditions (including the honeymoon) have grown up around honey wines. Strictly speaking, a wine made from honey alone is called mead. Wine that combines honey with apples is called cyser, whereas wine made from honey and any other (non-grape) fruit is called melomel. Wine made from honey with added herbs is called metheglin, and wine made from honey and grapes is pyment.

When making mead variants, the source and quality of the honey you use makes a difference in the taste of the finished product. The generic blended honeys in the supermarket are fine when the honey is primarily used as a source of sugar. If you are making mead, however, blended honey is useless because it has been pasteurized and homogenized until it is nothing but sugar. If the tastes and aromas of the honey will be important to the end product, use a single-source honey from a bee keeper. The nectar that the bees collect positively affects the mineral content and flavor of mead. Clover, alfalfa, orange blossom, wildflower, and mesquite are just a few of the types of honey available; in general, darker honeys impart stronger flavors. You can get single-source and wild honeys from a local bee keeper, order them over the Internet or, if you are ambitious, start keeping bees yourself[12] and create your own honey.

Cleanliness and Sanitation

Before I get into the details of making wine, I want to delve a bit into cleanliness and sanitation, as these are crucial for a successful outcome. You don't need a laboratory clean room or a level 3 hazmat facility to make wine. You can make wine in your kitchen or any room in your home. But you need to be attentive to detail. Everything that touches your must, wine, or wine-in-progress must be clean and sanitized.

---

12 In my opinion, I don't have enough experience to write a book on beekeeping, however I have found Kim Flottum's *The Backyard Beekeeper* to be very good. If you are interested in keeping bees for honey, I highly recommend it.

"Clean" simply means "free of visible dirt or contamination." Dish soap and water are adequate cleansers. Wine bottles, fermentation vessels, wine thief, plastic tubing, and hydrometer along with all utensils that touch your wine need to be cleaned. Sometimes, all that is needed is to add some soapy water and shake. Other times, as with carboys, you may need to use a special brush. For subsequent sanitation procedures to work, the surfaces must first be clean. Once they have been cleaned, they should be thoroughly rinsed.

To sanitize the equipment, all surfaces that will touch the must or wine should be rinsed or wiped down with a sanitizing sulfite solution. Don't rinse afterwards. For bottles, vessels, and carboys you can add a portion of the sulfite solution and swish it around thoroughly so that it contacts all surfaces, and then pour it back into your container of solution. For other utensils, soak paper towels in sanitizing solution and use those towels to wipe them down immediately before use.

Your hand siphon and tubing might look to pose a problem at first, but there is an easy technique for keeping them clean. For this technique, you need two clean plastic gallon jugs that were previously used for water. Put one with soapy water on your counter and the empty one on the floor. Now, use your siphon to pump the soapy water all the way up into the tube and through the tubing into the empty container on the floor. Then, switch the containers and repeat the process until the equipment is clean. Empty out the containers and rinse them thoroughly. Next, put the bottle of sanitizing solution on the counter, and siphon that into the empty container on the floor. Make sure to wipe down the outside of the equipment and tubing as well, as these may contact the wine.

Making the Wine Must

As noted previously, your wine must doesn't have to be made from a single source. You can use apples mixed with pears, carrots mixed with apples, juiced table grapes combined with bottled cherry juice, or whatever strikes your fancy. As long as you use good sanitation and technique, the results will be at least as good as most wines you can buy.

Some fruits are either highly acidic or highly tannic to such an extent that you wouldn't want to use their extracted juice exclusively to make wine because the results would be too sour or bitter. In those cases, only a portion of the must is made from that fruit, and the rest is made up from water or other juices.

What follows is a recipe table that indicates how many pounds of a given fruit to use in making a gallon of wine from that fruit, how much tannin to add to that

General Recipes for Making Wine from Common Fruits and Flowers

	Pounds	Fruit Preparation	Adjuncts	Pectic Enzyme	Tannin	Yeast Variety	Notes
Apple	10 lbs	Juice machine	½ lb raisins in straining bag	Double specified on container	¼ tsp	Red Star Pasteur Champagne	Use the juice but not the pulp
Blackberry	4 lbs	Crushed in straining bag	½ lb raisins in straining bag	As specified	None	Red Star Montrachet	
Blueberry	8 lbs	4 lbs juiced, 4 lbs crushed in straining bag	½ tsp vanilla and one stick cinnamon in bag	As specified	None	Lalvin ICV-D$_{254}$	
Cherry, Sweet	4 lbs	Juice machine, add pulp to straining bag	1 quart bottled cherry juice	As specified	None	Red Star Montrachet	Exclude the pits
Dandelion	5 cups flower heads	Put the flower petals (only!) in straining bag	1 lb raisins in straining bag	None	¼ tsp	Lalvin D$_{47}$	Ferment at under 70 degrees
Nectarine	3 lbs	Pitted and juiced in machine	½ lb raisins in straining bag	As specified	¼ tsp	Red Star Pasteur Champagne	
Peach	3 lbs	Pitted and juiced in machine	None	As specified	¼ tsp	Lalvin D$_{47}$	

General Recipes for Making Wine from Common Fruits and Flowers

	Pounds	Fruit Preparation	Adjuncts	Pectic Enzyme	Tannin	Yeast Variety	Notes
Pear	10 lbs	Juice machine	½ lb raisins in straining bag	Double specified on container	¼ tsp	Red Star Pasteur Champagne	Add one stick cinnamon to straining bag
Plum	5 lbs	Pitted and crushed in straining bag	1 lb raisins in straining bag	As specified	None	Red Star Montrachet	
Raspberry	4 lbs	Crushed in straining bag	None	As specified	None	Red Star Montrachet	
Rhubarb	4 lbs	Crushed in straining bag	1 lb strawberry in straining bag	As specified	¼ tsp	Red Star Pasteur Champagne	
Strawberry	4 lbs	Crushed in straining bag	½ lb raisins in straining bag	Double specified on container	¼ tsp	Lalvin ICV-D$_{254}$	Comes out straw-colored rather than red
Watermelon	10 lbs	Juice machine	1 lb raisins in straining bag	As specified	¼ tsp	Lalvin D$_{47}$	Peel the outer skin off, and juice the rind and fruit

wine per gallon, and any other adjuncts that I'd recommend. Any deficit in juice to make a gallon is made up with water.

The Primary Fermentation: Step-by-Step

1. **Start with fruit** juice obtained as described earlier in this chapter.
2. If needed, add enough water to the fruit juice to equal the amount of wine you wish to make. (It is helpful to add previously-measured amounts of water to your primary fermenter in advance and use a magic marker to mark gallons and quarts on the outside of the vessel for easy reference.) I use bottled water because my well water is sub-optimal, but if you have good water where you live, tap water is fine. Don't worry about whether or not your water is chlorinated, because the Campden tablets we'll be adding later serve to remove chlorine from the water.
3. Use your hydrometer to measure the specific gravity (SG) of the must. You are aiming for an SG of between 1.085 and 1.110, but in all likelihood your must measures much lower. Add the required amount of sugar or honey to your must. This will slightly increase the volume of your must, but that's fine.
4. Use your acid testing kit to test the acidity of your must. If needed, add acid. Try to use the specific acid (or acid blend) that will best enhance the character of the fruit. For example, malic acid will enhance apples and pears whereas citric acid will enhance watermelons and tartaric acid will enhance grapes. If you are in doubt, use an acid blend made up of equal parts of the three acids.
5. Add one teaspoon of yeast energizer for each gallon of must.
6. Add pectic enzyme as directed on the container, or double the amount if the recipe table specifies doing so.
7. Add tannin as appropriate for the fruit being used. (This is described in the accompanying recipe table.)
8. Crush and add one Campden tablet dissolved in a bit of juice per gallon of must. Vigorously stir the must.
9. Cover your primary fermenter, and plug the hole with a bit of cotton ball to keep foreign objects out. Wait 24 hours.
10. Vigorously stir the must to oxygenate. Once movement ceases, sprinkle yeast from the packet over the surface of the must. Do not stir.
11. Place the cover on the fermenter, and plug the hole loosely with a bit of cotton ball.

12. Allow to sit for a week. During this time, you should smell the fermentation. Also, it may foam heavily and come out through the hole in the lid. If it does, clean up on the outside and insert a new cotton ball. After the week, replace the cotton ball with an air lock filled with sanitizing solution.

13. Allow to sit for another week or two, until the air lock only "bloops" once every few seconds. This marks the end of the primary fermentation phase. Once the primary fermentation phase has ended, rack as soon as possible. If it sets for more than a couple of days, the dregs at the bottom (known as lees) will impart bad flavors to your wine.

Your First Racking: Step-by-Step

1. **A day before** you plan to rack, move your primary fermenter to a table or counter top. (By doing this a day in advance, you give any sludge stirred up by movement a chance to settle.) Put a wedge, book, block of wood, or something else from 1" to 2" high underneath the fermenter on the edge that is away from you. This will allow you to sacrifice the smallest amount of wine possible with the lees.

2. Clean and sanitize your racking tube, tubing, and your secondary fermenter and get your rubber stopper and fresh air lock ready. Put your secondary fermenter on the floor in front of the primary fermenter, and then carefully remove the lid from the primary fermenter, creating as little disturbance as possible. Put a bit of the wine in a sanitized glass, and dissolve one crushed Campden tablet per gallon, then add it back to the wine.

3. Put the plastic tubing from your racking setup in the secondary fermenter and the racking tube in the primary fermenter. Keeping the racking tube well above the sedi-

≫ Racking requires a bit of coordination but with practice it comes easily.

ment, pump it gently to get it started. It may take a couple of tries. Gently lower the racking tube as the liquid level diminishes. Watch the liquid in the tube very carefully. The second it starts sucking sediment, raise the racking tube up to break the suction.

4. Place the rubber stopper with an air lock filled with sanitizing solution in the secondary fermenter. Using a carboy handle if necessary, move the secondary fermenter to a location out of sunlight where even temperatures are maintained.

5. Immediately clean and sanitize your primary fermenter, racking tube, and tubing and stow them away. If you don't clean them immediately they will likely be ruined.

The Secondary Fermentation: Step-by-Step

1. **Wait. And wait.** Then wait some more. Patience, I have been told, is a virtue. After your first racking from the primary to secondary fermenter, the yeast will lag for a couple of days while it tries to catch up. Allow the secondary fermenter to sit unmolested until: the wine starts to become clear, you have more than a dusting of sediment at the bottom of the container, or the air lock only operates once every couple of minutes. This will likely take about a month. Because the secondary fermenter sits for so long, don't forget to check the airlock periodically and top up the sanitizing solution in it so it doesn't evaporate and leave your wine vulnerable.

2. The day before racking, sit the secondary fermenter up on a table or counter, and tilt it using a book or other wedge, as before.

3. Boil some water in a pot on the stove and allow it to cool to room temperature while covered. Clean and sanitize

 The secondary fermentation in this vessel is almost complete.

another fermentation vessel as well as the racking tube and tubing. Place the sanitized vessel on the floor in front of the secondary fermenter, put the plastic tubing in the vessel, and put the racking tube in the secondary fermenter. Operate the racking tube and transfer the wine from the old vessel to the new. Add one Campden tablet dissolved in wine for each gallon of wine.

4. Likely, there is an air space between the wine and the top of the new vessel. This will expose too much surface area of the wine to oxygen and potential infections. Pour in the sterilized water until the wine is just up to the neck.

5. Clean the rubber stopper and airlock, sanitize them, re-fill the airlock with sanitizing solution, and install them on the new secondary fermenter. Put the secondary fermenter in a location without sunlight and with even temperature.

6. Thoroughly clean and sanitize the empty secondary fermenter, the racking tube and the tubing.

7. Now it is a waiting game. Over weeks and months your wine will ultimately cease to ferment, and the haze within the wine will settle onto the bottom of the container. Keep an eye on the wine. Anytime a substantial sediment develops, rack the wine again and top up with sterilized water. Make sure the airlock doesn't go dry and permit foreign organisms to enter. Depending on ingredients, you may not have to rack again or you could have to rack one to three more times.

8. Once the wine has gone at least three months without requiring racking and is crystal clear, it is ready for bottling. You can allow it to age in the fermenter as long as you'd like—even several years.

Bottling Your Wine:

1. **Gather, clean, and sanitize** the wine bottles that will be accepting the wine. You will need five bottles per gallon of wine. Boil an equal number of corks for 15 minutes, and then allow them to sit in a covered pot. Clean and sanitize your corker.

2. Rack the wine, but do not add water to top off on this final racking. Add one Campden tablet per gallon by crushing the tablet and dissolving in a bit of wine and then adding that wine into the new vessel.

3. If you want to add potassium sorbate to prevent re-fermentation, dissolve that also in the wine and add back into the new vessel. Use ¼ tsp per gallon of wine.

4. Place the secondary fermenter holding the wine on a table or counter top, tilting as before.

» Bottled blueberry wine. Magnificent!

5. Arrange the bottles on a towel on the floor in front of the fermenter and install the plastic hose clamp on the plastic tubing so you can turn off the flow of wine when switching from bottle to bottle.

6. Put the plastic tubing in a bottle and the racking tube in the wine, pump the racking tube and start the flow of wine, pulling the tube higher in the bottle as the level of the wine increases. Stop the flow of wine when it is about half an inch into the bottom of the neck of the wine bottle. Put the tube in the next bottle and repeat the process until you run out of wine or there are no more bottles to fill.

7. One at a time, place the bottles on a solid floor, and use your corker to install one of the sterilized corks. Only work with one bottle at a time. As each bottle is corked, set it out of the way. Repeat until all of the bottles are corked.

8. Clean and sanitize your fermenting vessels, racking tube, and plastic tubing.

9. Now you can make and apply wine labels if you wish.

10. Enjoy at your leisure or give as a gift.

Creating Your Own Recipes

Most wine books are full of recipes. I'm sure they are useful, but I think it is more useful to understand how they are created so you can make your own recipes based upon what you have available. One of the reasons I went into so much detail in the chapter on the science of making wine is so you'd have all the tools you need to be self-sufficient.

When making a new wine for the first time, it is best to make it in a one-gallon batch. Overall, larger batches tend to come out better because they are less susceptible to temperature fluctuations, but I've made plenty of excellent wines in one-gallon batches, and the smaller amount may be less intimidating to a beginner.

In addition, even the cheapest wine ingredients are expensive at the scale of a five-gallon batch, so making a one-gallon batch is also best for experimental recipes. Reserve the larger five-gallon batches for tested recipes you know you'll be making for long-term storage or gifts. Some wine makers never make batches larger than a gallon, and they are quite happy with their results.

So let's look at the first critical decisions involved in making a wine recipe.

- Dominant and any secondary or tertiary fruits or flavors. This decision is purely aesthetic, and if you are someone with chronically bad taste you might consider consulting a friend or loved one for guidance. For example, I might want to make a wine with apple dominant, mulling spice secondary and honey tertiary. It would be rather daring as there would be little residual sweetness to balance the spice, so instead I'll make the apple dominant, honey secondary and mulling spice tertiary. Another example would be a wine with sweet cherry dominant and concord grape secondary.
- Check the tannin levels of each fruit. Any fruit that is high in tannin cannot be more than half of your must. If you want the fruit to be dominant anyway, you'll have to choose something unassertive such as generic white grape juice as a secondary. For example, if I want to make a wine with blueberry dominant, because blueberry is highly tannic, I am limited to four pounds of blueberries per gallon and will need to use adjuncts that won't overshadow the blueberry, such as white grape juice to make up the difference in volume.
- Check the acid levels of each fruit. Though the tables I've included are not a substitute for actual measurement, you can use them to get an idea that lemons are too acidic to constitute a major proportion of a wine must and that watermelon would need acid added. If the level of acid in the fruit is greater than 9g/L, then the quantity of that fruit should be limited to avoid excess acidity.
- Sugar level. Get a general idea of how much sugar will be in the major fruit ingredients, and how much will need to be added. Also, decide what form that sugar will take.
- Spicing. Look at the earlier tables when deciding how much spicing to add, if any.
- Pectic enzyme. Other than straight meads, all recipes will need pectic enzyme. Review the tables to see if the amount used is according to package directions or needs to be doubled.
- Yeast. Decide what type of yeast to use based upon its characteristics, and include yeast energizer in the amount recommended on the package.

One Gallon Example: Cherry Wine

Few can resist the idea of cherry wine, so it's a good start for a recipe. I've made cherry wine during the winter months and used a combination of bottled cherry juice, frozen cherries, and other adjuncts.

- I would like my dominant flavor to be cherry, my secondary flavor to be grape, and my tertiary flavor to be vanilla.
- Checking the tables, I see that cherries are high in tannin so they can't form more than half of my must, and they are high in acid too. So I will use one quart of bottled organic 100% black cherry juice, and one ten-ounce package of frozen sweet cherries, and make up the difference with organic white grape juice. Because less than half the recipe will have the higher tannins of cherry juice and the remainder of the recipe will come from white grape juice, tannin is likely to be a bit on the low side, so I'll add a quarter teaspoon of tannin.
- The two primary acids in cherries are malic and citric in that order, so if an acid test shows more acid is needed, I'll add a 2:1 mixture of malic and citric acids.
- Because I'm using bottled juices and frozen cherries with nutritional labels, I can add up how much sugar would be in the gallon: 200g from cherry juice, 42g from frozen cherries, and 480g from grape juice for a total of 722 grams. Converted to ounces that is 25.5 ounces. Looking at the hydrometer tables, I see that I need 31.6 ounces of sugar for a starting gravity of 1.090, so I figure I'll need about six ounces of sugar, though I will test with a hydrometer to be sure.
- Looking at the spicing table, and not wanting to overpower the cherry, I am including one vanilla bean in the primary fermenter.
- A lot of yeasts are available, but Montrachet looks like a really good match for what we're making.

So the final recipe looks like this:

Winter Cherry Wine

1 quart organic black cherry juice
1 ten-ounce package of frozen sweet cherries
3 quarts organic white grape juice
6-8 ounces of sugar, based on hydrometer test
1 vanilla bean
¼ tsp yeast energizer
¼ tsp tannin
2:1 blend of malic/citric acids as needed
1 packet Montrachet yeast

6

Advanced Techniques in Winemaking

Some of the techniques in this chapter truly are "advanced" in terms of difficulty or equipment, but most are merely extensions of what you already know that will help you produce wines with different characteristics.

Making Wine Without Sulfites

In the previous chapter I oriented procedures around the use of sulfites, because the use of sulfites makes it easier to produce a solid product. Even before modern times, sulfites were employed by burning sulfur to create sulfur dioxide gas to purify musts, and many yeasts generate their own sulfites during primary fermentation. There is no such thing as a sulfite-free wine. The best that can be managed is to make wine without adding them.

If you don't want sulfites added to your wine, here are some tips:

- Clean all equipment using scalding (140+ degree) water. Water this hot is, in fact, scalding. Use caution.
- Though it decreases the quality of the result somewhat, you can pasteurize your must by heating to 150 degrees and holding it there for ½ hour before putting it in your primary fermenter. Add bottled water to make up any difference in volume. Make sure the temperature has dropped below 80 before adding pectic enzyme or yeast. Musts that have been pasteurized often create very hazy wines, so don't do this with fruits that require a double dose of pectic enzyme.
- When using unpasteurized must, your yeast must out-compete all other yeasts and bacteria. So instead of sprinkling the yeast on top of the must, get your yeast ready two days in advance by sprinkling the yeast into a pint jar containing bottled apple juice mixed with a pinch of yeast energizer and then cap it with several layers of cheesecloth. Set it in a dark place at 60-70 degrees. When you add this yeast to your must, just pour it in smoothly and do not stir. Now your yeast has a head start so it can out-compete the wild yeasts and bacteria on the fruit.

Advanced Testing of Sulfite Levels

The sulfite test kits you can buy through wine hobby suppliers are fine for lightly colored wines but inaccurate when testing heavily colored wines because the phenolic coloring compounds in the wine take up some of the testing reagent. Errors can be as high as 20ppm. In an earlier chapter, I advised that you could "guesstimate" by subtracting 10ppm from the test results. If you aren't comfortable with guesstimates, you can do your own testing. The following procedures work on the principle of calculating total sulfite, calculating the error, and then subtracting the error from the total. You will be working with sulfuric acid in this procedure; safety goggles and a lab apron are required.

Equipment

50 ml beaker
3 ml syringe (no needle needed)
5 ml syringe (no needle needed)

10 ml graduated cylinder
250 ml volumetric flask (only needed if preparing your own iodine solution)
Scale (only needed if preparing your own iodine solution)

Chemicals

Distilled Water
.02N Iodine solution[13]
1% Starch solution
25% Sulfuric Acid solution
3% Hydrogen peroxide solution

Clean all equipment with distilled water. Put a 20ml sample of wine in the 50ml beaker. Add 5ml of starch solution, swirl to mix, then add 5ml of sulfuric acid solution to the sample and swirl to mix again. Fill the 3ml syringe with iodine solution. Add iodine solution a little at a time, swirling after each addition, until a distinct color change (it will be dark blue) that remains for several seconds occurs. The measured amount of sulfite in ppm is:

(3ml–reading on syringe) x 32

Write this number down as we'll use it twice in the next part of the procedure. Now we need to measure the error. To do this, we'll need to remove the free sulfite from the solution. The 3% hydrogen peroxide solution is way too strong for a sample this small, so make some 0.012% solution by adding 1ml of hydrogen peroxide solution to 250ml volumetric flask and adding enough distilled water to meet the 250ml mark. Clean all equipment with distilled water. Put a 20ml sample of wine in the 50 ml beaker. Add an amount of your prepared hydrogen peroxide solution equal to 0.14 ml for every ppm detected in the earlier procedure. So if the first procedure gave a result of 57ppm, add 0.14 x 57 or 8 ml of hydrogen peroxide solution. Swirl to mix and wait a few minutes. Add 5ml of starch solution and 5ml

---

13 This can be purchased from a lab supply company or made. To make it, mix .63g of iodine crystals and 1.3g of potassium iodide crystals together, place in a 250ml volumetric flask, and fill to 250ml with distilled water. Due to methamphetamine labs, iodine has been regulated since 2007 as a precursor. You can still buy elemental iodine, but your name goes on a list. Your name won't go on a list if you buy ready-made .02N iodine solution, but it is pretty expensive. If you aren't running a methamphetamine lab, don't worry about your name being on a list.

of sulfuric acid solution to the sample. Fill the 3ml syringe with iodine solution. Add iodine solution a little at a time, swirling after each addition, until a distinct color change (it will be dark blue) that remains for several seconds occurs. The measured amount of error in ppm is:

(3ml−reading on syringe) x 32

The corrected amount of sulfite in the wine is: (measured sulfite)−(measured error) ppm.

Malolactic Fermentation

Most wine musts will contain some lactic bacteria. These are inhibited by sulfite levels greater than 20 mg/L, by alcohol concentrations of greater than 14%, low temperatures, and active yeast. If you have ever been quite certain that secondary fermentation has completed but later found your bottled wine to be slightly carbonated, it is likely that malolactic fermentation occurred spontaneously. Sometimes it even occurs simultaneously with alcoholic fermentation and you never notice.

Commercial wineries subject a large proportion of their red wines and a lesser proportion of their white wines to deliberate malolactic fermentation for a variety of reasons. The most obvious reason is that malolactic fermentation changes sharp malic acid to smooth lactic acid. It raises the pH and reduces the acidity slightly. The formation of lactic acid in the presence of ethanol also allows the creation of ethyl lactate, an ester that gives wines a fruity character. In addition, bacteria will produce diethyl succinate, another fruity ester along with diacetyl and other flavor compounds. Malolactic fermentation also serves to make wine more self-preserving by consuming pentoses and hexoses[14] not used by yeast, as well as malic acid that would otherwise serve as food for other bacteria. Malolactic bacteria secrete bacteriocins that inhibit the growth of other bacteria, which makes the wine more microbiologically stable. Finally, by deliberately conducting a malolactic fermentation, you can be certain that one won't occur spontaneously in the bottle.

There are a number of malolactic cultures available. Some are single species (usually *Oenococcus oenii*) and others contain a mix of species. Only *Oenococcus*

---

14 Pentoses and Hexoses are five-and six-carbon sugars, respectively.

oenii can work at pH values lower than 3.5, so all cultures contain at least that one bacteria.

Malolactic bacteria and wine yeast are often incompatible as one will inhibit the other. Therefore, malolactic culture is best introduced after the secondary alcoholic fermentation is complete unless the wine has a final alcohol level exceeding 14%. Though different products will require slightly different procedures and I based the following on the use of Lalvin's malolactic culture, it's a good guideline for malolactic fermentations generally.

- Wait until secondary fermentation has completed. It can be considered complete after the wine sits for three months without dropping any precipitate at the bottom of the fermenter and it is clear.
- About a week before adding the malolactic cultures, move the wine to an area with a temperature between 64 and 75 degrees. It must remain at this temperature throughout the fermentation.
- Rack the wine into a new, clean secondary fermenter. Do NOT add sulfite in this racking!
- If you have a sulfite test kit, you can test that the sulfite is under 20 ppm. If it is above 20 ppm, you can reduce it to that level via the addition of hydrogen peroxide as described in the chapter on wine chemistry.
- Add the packet of malolactic culture directly to the wine.
- Fit with an airlock. (The fermentation generates carbon dioxide.)
- Malolactic fermentations complete in one to three months, but because they proceed so slowly, it is hard to gauge. You could assess the progress using commercially available paper chromatography kits for malic acid detection, but these cost from $50 to $200 and their shelf life is only a few months. Unless you are making a LOT of wine, this doesn't make a lot of sense.
- After three months, rack the wine into a clean secondary fermenter, adding 50ppm sulfite to the must (one Campden tablet per gallon). This will inhibit any malolactic bacteria that remain. Allow to age for at least another month, then bottle.

Fortification

Fortification is the addition of distilled spirits to wine in order to increase its alcohol content. Wines are fortified for three reasons: to bring them to an alcohol level suitable for the development of what is known as a sherry flor; to arrest fermentation before all the sugar has been consumed so it retains sweetness; and to make a more biologically stable wine for purposes of long storage and shipping. Being distilled from wine, brandy is the most common spirit used for fortification.

Sherry wines are fermented to dryness, their alcohol content is assessed, and then they are mixed with brandy to reach an alcohol level of between 14.5% and 16%. This range is conducive to allowing the specific strain of yeast used in sherries to develop a cap known as a flor that protects the wine from further oxidation and promotes aldehyde production. If the wines were good but unexceptional, they are made into oloroso sherry by fortifying to a level exceeding 16% so that a flor isn't formed. If they were really bad, they are set aside to make vinegar or brandy.

Port wines are fortified to an amount of alcohol around 20% before the primary fermentation has completed. The high level of alcohol from fortification stops the fermentation very quickly so that high sugar levels remain.

Because distilled spirits are expensive (why would you use the cheap stuff?), the financially practical batch size for fortification will likely be limited to a gallon or so. Even though fortification will stop the yeast fermentation, the wine will continue to undergo small changes from aging for as long as forty years.

The first step in fortification is to assess the level of alcohol in the wine. For wines fermented to dryness, this is easy—you can just use the potential alcohol corresponding to the original hydrometer reading of the must. For wines that will retain sweetness, you will take a hydrometer reading every day or two during the primary fermentation and perform the fortification at a point corresponding to the degree of sweetness you want to retain. In general, you'll want to do this at a reading between 1.010 for slightly sweet to 1.040 for very sweet. The alcohol level level can be determined by subtracting the potential alcohol at the point of fortification from the potential alcohol at the start of fermentation.

For example, if the original specific gravity of my must was 1.093 (12.8% potential alcohol) and I perform my fortification when the must reaches 1.027 (3.7% potential alcohol), the level of alcohol is 12.8% - 3.7% or 9.1%.

Once you know how much alcohol the wine already contains, you must decide how much alcohol you want it to contain after fortification. If you are wanting a sherry flor, you'll want 15%, but if you are making a sweet wine, you'll want 20%.

Then you need to calculate how much distilled spirit and how much wine to add to make a gallon at the desired strength. Though there are tables for this, algebra is the most flexible tool, and the equation is easy enough.

A = Percentage of Final Wine as Alcohol expressed as a decimal (e.g. 20% = .2)
B = Percentage Alcohol of Starting Wine expressed as a decimal

C = Percentage Wine of Distilled Spirit expressed as a decimal

D = Size of final batch in ounces

X = Ounces of Wine for the batch

X = D (A–C)/(B–C)

So, if I want a wine with 20% alcohol, my starting wine is 9.1% alcohol and I am fortifying with brandy that is 40% alcohol to make a batch of one gallon (128 ounces):

X = 128(.20 - .40)/(.091 - .40) = 82.8 rounded to 83 ounces.

So now I know I will use 83 ounces of wine and 128–83 or 45 ounces of brandy to make a gallon of my sweet fortified wine.

The procedure is straightforward. You put 45 ounces of brandy in a secondary fermenter, rack wine into the fermenter until you have a gallon, and then top it with an airlock. If you are making a sweet wine you'll need to rack it in another week or so, and then treat it like any other wine. If your starting wine was already completely fermented, you can wait a month before racking and then treat it like any other wine.

Oak Aging

Oak aging is traditional in wines, especially red wines. There's no question that the traditional aging of wines in oak barrels alters the flavor through the extraction of a variety of compounds from the oak. However, oak barrels are very expensive and require considerable care. Studies show that using a neutral container (such as glass) and adding toasted oak chips will impart the same compounds as aging in a barrel.[15]

Oak chips are available from both American and French oak, and in a variety of toasting levels. French oak imparts more tannin and spice notes, whereas American oak imparts more vanilla and sweet notes. Toasting oak cubes of either sort makes some of the compounds such as vanillin more available but also imparts a more charred character to the wine, especially at the highest levels of toasting.

Oak chips impart their character quickly at first, and then more slowly over time. Once they have been added, they can be removed and the compounds they add will remain and continue to work throughout the aging process. Though they

---

15 A. Bautista-Ortín, A. Lencina, M. Cano-López, F. Pardo-Mínguez, J. López-Roca and E. Gómez-Plaza (2009), "The use of oak chips during the aging of a red wine in stainless steel tanks or used barrels: effect of the contact time and size of the oak chips on aroma compounds."

can be retained in the fermenter for as long as nine months, little is gained by keeping them for longer than a month.

Before being added to the fermenter, the chips need to be sanitized so they don't infect your wine with something nasty. All you need to do is fill a quart canning jar with water, add a quarter teaspoon of potassium metabisulfite powder, put your oak chips in the water, and then put on the lid. After twenty four hours, the chips can be added to the wine by simply dumping them into the wine in the secondary fermenter. They won't fit through the racking tube, so they'll be removed at the next racking. Just rinse them out of the fermenter when you clean it.

Solera Aging

A solera is a grouping of containers (usually barrels) used to accomplish a unique blended aging technique in the production of certain vinegars, spirits, and wines. On a commercial scale, a solera is a substantial investment, but for an amateur with containers not exceeding five gallons, the larger concern in space. Though you can technically use this technique with as few as two containers and with no upper limit, I am going to describe it using three.

Label three containers as A, B, and C, filling them all with wine. After the wine has aged a year, withdraw and bottle half of the contents of container C. Refill container C from container B, refill container B from container A and use new wine to refill container A. Do the same thing every year, and over time the average age of the wine bottled from container C will approach five years, even though you are bottling wine from it every year.

The technique of solera aging can be employed at home on a smaller scale using simple glass fermenters.

The average age increases with the number of containers and with bottling smaller portions according to the following formula:

Average Age = (Number of Containers–Fraction of Container Used)/Fraction of Container Used

So if you started with three five-gallon containers and only bottled one gallon (.2 of a five-gallon container) a year, the average age would approach (3 - .2)/.2 or 14 years. If you used four containers instead of three and drew off half of a container each year, the average age would converge upon seven years.

If you have a particular type or style of wine that you like and that requires substantial aging, setting up a solera of three containers can be an easy way of having your cake and eating it too, where over time you can consistently produce well-aged wine in small quantity every year.

Of course, this exact same approach is used with the higher quality balsamic vinegars, many of which have average ages exceeding a decade. Because vinegar is consumed in smaller quantities than wine, it is entirely practical to set up an inexpensive vinegar solera using five one-gallon containers. If you drew off and replenished a half gallon annually, the aging would converge upon nine years.

PART III
Beer from Seed to Glass

7

Overview of Beer Making

Drinking beer, according to television commercials, will make you attractive and successful. If you are reading this book, you are probably dubious about such implicit marketing claims, but you may well enjoy beer because of its unique and refreshing taste.

Beer has been around for thousands of years,[16] is the world's most popular alcoholic beverage and the third most consumed drink in the world, behind only water and tea. The price of domestic beer ranges from $7.50/gallon to $11.00/gallon when purchased by the case. Per capita annual beer consumption

---

16 *New York Times,* 11/5/1992 "Jar in Iranian Ruins Betrays Beer Drinkers of 3500 B.C."

is only 20.5 gallons in Utah, but an impressive 43.5 gallons in Montana. Considering that many people don't like beer, it's safe to say that those who do drink it, drink more than the per-capita statistics would indicate. Likewise, not all of that beer is purchased by the case, so it is far more expensive than the case prices. A household with two adults can easily spend hundreds of dollars on beer every year.

Consequently cost is a good reason why making your own beer is well worth consideration. Though the ingredients for making your own beer are not free, they are about a quarter of the cost of buying beer at the store. Furthermore, beer is a very interesting beverage and well worth learning how to make yourself. Just as learning to grow your own vegetables and make your own wine can offer the benefits of a product far superior to what can be purchased at any price, once you get the hang of making beer you can routinely have beer of such high quality that buying it, if possible at all, would be cost-prohibitive.

Most books on home brewing are divided into beginner, intermediate, and advanced techniques. Beginner approaches use malt extracts, and intermediate methods combine malt extracts with infusions of grains that don't require mashing. Advanced methods rely strictly on malted grains and mashing.

The approach covered in this book relies upon malted grains and mashing, and would thus be classified as advanced. I will admit there are a lot of variables to be considered when making beer this way. However, the methods really aren't complicated and the minutiae may make the difference between a good beer and a great beer, and if you mess up a bit, you'll still end up with good beer 99% of the time.

In addition, brewing with all-grain methods is justified by virtue of the fact that it lends itself to greater self-sufficiency than being reliant on malt extract and kits. Malted grains cost a lot less. In fact, most beer kits cost between $8 and $15 per gallon, so it wouldn't save you any money at all, whereas beer from malted grain can easily cost as little as $4 per gallon—even less if you grow and malt your own grain.

The cost of equipment to get started is, to be honest, daunting. If you had to buy everything needed right off the bat, it would cost $300-$400. But if you are already making wine, you can use the same fermenters, testing equipment, and many of the same additives. Likewise, if you already make cheese, you use the large spoon and digital thermometer needed to make beer. So if you are already undertaking these aspects of self-sufficiency, making all-grain beer will only require a few items you don't already have.

Even with the worst case scenario of having to buy everything to make beer, as the average household will consume as much as eighty gallons of beer annually you will still break even on the cost within a year, and save hundreds of dollars going forward, since most of the equipment you use will easily last a lifetime.

In summary, my arguments in favor of making your own beer come down to the fact that you will save money in the long run so you can use that money differently, you will have a superior product, you can choose to use organic ingredients thereby reducing your exposure to unintended poisons, and you'll have the confidence that comes from controlling your own supply chain.

Types of Beer

Because beer is so widely consumed, there are dozens of variants. For our purposes I'm going to divide beer into categories based upon the type of yeast that is used, the type of malt that is used, and the type of conditioning it undergoes.

Ale is beer that is fermented using a top-fermenting yeast. Ale is fermented at room temperatures between 60 and 75 degrees. Though today ale is usually flavored with hops, hundreds of years ago ale was originally flavored with a mixture of herbs called gruit. A common gruit mix would include mugwort, yarrow, and rosemary. There are many variations of ale based upon the degree of roasting of the malts used to make it. Porter and stout are also variants of ale.

Lager is beer that is fermented with a bottom-fermenting yeast and undergoes primary fermentation at temperatures between 45 and 54 degrees followed by storage at 32 to 39 degrees. Storing a beer at that temperature range is called lagering. What you will most likely encounter at the ball park, and the type of beer most often consumed, is a pale lager.

There are many variations on these themes, both in terms of technique and ingredients. Examples include altbier, which is top-fermented at room temperature like ale, and then lagered, and California common, which uses a special bottom fermenting yeast that works well at room temperature.

Once you know the ingredients and techniques, it's your beer, so you can create it any way you wish. That's one of the joys of home brewing!

Essentials of Beer and Brewing

What is beer? I am going to skip the dictionary definition that includes all sorts of stuff of little interest and give the American definition: Beer is a fermented alcoholic beverage flavored with hops and made from malted grains such as barley and wheat.

Grains are starchy seeds. In their dry form, yeasts can't digest them because they are all starch and no sugar. However, many grains in the process of sprouting produce enzymes that will turn the starches into sugars. Malting grain is little more than inducing it to sprout, halting the germination process at the right time by heating the grain enough to kill it but not enough to damage the enzymes, and then drying and cracking the sprouted grains. Barley and wheat are the most commonly malted because their enzymes are the most efficient for converting starches to sugars.

The cracked malted grains are mixed with water in a process known as mashing. Because the enzymes that convert starch to sugar work best at certain temperatures, the mash is held at those temperatures until the process of converting starch to sugar has completed. The liquid that will become beer is called a wort. Then, the soluble sugars and other nutrients are washed out of the mash into the boiling kettle in a process known as sparging. Sparging serves two purposes: it helps obtain the sugars resulting from enzymatic conversion of starch, and it uses the grain bed as a filter to remove solids. The sugary water is then boiled. This stops enzymatic activity and sterilizes the wort. Hops are added at different points in the boiling cycle depending on their type and purpose.

Toward the latter part of the boiling period, a device called a wort chiller is put in the boiling wort. When the boiling is over, the wort chiller is used to cool the wort to a proper range for fermentation as quickly as possible. Once it is cooled, specific gravity readings are taken, it is put into a sanitized primary fermenter, and the yeast is pitched. Once the yeast has done its job in anywhere from two days to two weeks, the beer is racked off the trub (spent yeast and condensed solids) into a priming bucket containing a bit of sugar. The beer is then put into sterile bottles and capped. The bottled beer is ready after a couple of weeks and will develop improved flavor for up to three or four months.

This is really the essence of making beer. In principle it is fairly simple though it may be logistically complex due to the problems associated with handling hot or boiling wort in quantities as large as five gallons. You can, incidentally, make beer in smaller quantities. Fortunately, unlike the ingredients of wine which tend to be pretty expensive for large quantities, the grains used in making beer are very budget-friendly. So the primary reason for making beer in smaller quantities would be to avoid the expense of large brew pots and the means of heating them. The way I have it figured, if you want to, you can start with one gallon batches and then use the money you save to afford the couple of items you'll need for larger batches.

⊗ A simple beverage cooler makes an excellent mash tun.

Mash Tun

You will see the word "tun" used frequently in brewing literature. It is an antiquated term referring to wooden casks used in brewing back before our great-great-grandparents were born. In modern usage it just means "container dedicated to some beer-brewing task." Because of the long and tradition-laden history of brewing, people hold onto and re-task older terminology. A lot of the terms are also Germanic, and this is because of the impressive traditional history of brewing in Germany.

Mashing is a word used to convey holding a mixture of malted grain and water at certain temperatures until the enzymes have converted the starches to sugar. There are different approaches to mashing. In some approaches, it is done at different temperatures (called a step mash) and in others it is done at a single temperature only.

If you are using a single temperature mash, which is what I describe in this book, the most important thing is that it holds heat. For this purpose, a five-gallon beverage cooler is the least expensive option.

If you are using multiple temperatures, you could certainly still use a beverage cooler by doing calculated additions of water at a higher temperature, but your best bet would be a large pot because you can add heat to a pot easily. Because mashing doesn't use the full volume of wort, you can use a smaller pot than you'll need for a brew pot, or you can simply use your brew pot. If you are using pots, you will need a gas burner for adding heat. An electric burner simply isn't up to the task.

Lautering Tun

If you use a beverage cooler as a mash tun, you can fit it with a manifold that will allow it to work as a lautering tun as well, or you can make one easily from a couple of buckets fitted inside each other. Lautering is the process of extracting all of the sugars from the mashed grain while allowing that grain to serve as a filter to trap the larger solid particles. A lautering tun contains either a manifold or a

⟪ Don't spend a lot of money on a lautering tun when there are many ways to make it yourself.

false bottom with hundreds of small holes to hold the grain in place while allowing the water that runs through the grain to exit. Lautering can be accomplished via a continuous process in which water is added to the top of the tun as quickly as it is removed (called sparging), or it can be done in a batch process in which all of the sparging water is added at once and then slowly drained. When done using a continuous process, a manifold is used on top of the grain as well so that the water is evenly distributed around and through the grain, so as much sugar as possible is extracted.

The manifolds are actually very simple plumbing implements made from either copper or CPVC piping. Both of these materials are used for hot and cold water in residences and are perfectly safe. Because they aren't carrying any pressure when used in a lautering tun, the fittings require no cement or solder. You can simply fit them by hand.

Likely the easiest and cheapest way of making a lautering tun is to use a priming bucket with a spigot to catch the liquid, and a matching bucket inserted into it whose bottom has been drilled with hundreds of ⅛" holes to hold the grain for lautering.

I use a combined method in which I use small two-gallon buckets. The one on the bottom contains a spigot and the one that is inserted has the bottom drilled with holes for holding and filtering the grain. I have made a manifold that sits on top of the bucket, and pour wort into a funnel on top of the manifold as needed.

You can also buy these for $200 or $300, but I don't see much point in doing so as they can be readily manufactured at home at low cost.

Brew Pot(s)

You need a pot for boiling the full quantity of your wort. If you are making five-gallon batches, then you need a pot that will hold seven gallons. That is because your wort will lose volume to evaporation while being boiled, and you don't want your pot boiling over and making a mess. You can use this same pot for heating water for mashing and lautering, or you can use a separate smaller pot for these tasks. Though most use stainless steel for brewing, stainless steel pots of this size are very expensive—over $100 and sometimes approaching $200.

Enamelware, such as is typically found in water bath canners, is much more affordable and perfectly serviceable as long as it isn't chipped. (Any chips will make your beer taste like iron.) Aluminum pots are unpopular because many people think the aluminum will give them mental diseases, but if used properly (i.e. for heating and boiling rather than for storage) they are a better choice than either stainless steel or enamelware because they are both affordable and provide better heat conduction and control. One trick with aluminum is to leave the hazy patina intact instead of scrubbing it off. The hazy patina is an oxidation layer that protects the aluminum from what you are boiling, and protects what you are boiling from the aluminum.

I have found that seven-gallon turkey fryer pots are perfectly sized, and can cost under $40 even in a stainless steel version. That is what I use.

Propane Burner

Making beer will tie up your kitchen for at least eight hours. This may or may not be acceptable in your house. I'm lucky in that my wife doesn't mind because I thoroughly clean the kitchen after I'm done.

Unfortunately, some electric stoves aren't powerful enough to boil five or six gallons of wort. The solution to this problem is a standalone propane burner. They cost about $55 to $95, and they use the same size propane tank as a gas grill. Once you have one of these, you'll find it handy for all sorts of things, such as heating the water used for scalding chickens, frying turkeys and so forth. These burners are for outdoor use only. If you try to use them in a garage they will almost certainly start a fire, but the carbon monoxide may get you first. So outside only!

You can use an outdoor propane burner if your kitchen stove is insufficient or off limits.

Wort Chiller

Once the wort is boiled, it needs to be cooled to fermenting temperature quickly. Otherwise, spoilage organisms will have too much time to gain a foothold and the beer will spoil. A wort chiller is a simple coil of copper tubing fitted with hoses on either end for cold tap water to run through. The chiller is immersed

◈ A wort chiller cools your wort quickly to reduce the risk of contamination.

in the wort for the last fifteen minutes of boiling so it is sterilized, and once the boiling is done, run cold water through it. This removes heat much more rapidly.

If you don't have a wort chiller or don't want to go to the expense, you can chill the wort in the brew pot more quickly by putting it in a bathtub filled with cold water, and running cold water down the sides if you can do so without getting the water into your pot and contaminating the wort. This will take about 35 minutes, so it isn't as good as a wort chiller, but it is better than allowing for the several hours it would take the wort to cool sufficiently on its own.

Primary Fermenter

The primary fermenter is a large plastic bucket made of food-grade plastic. It is sized at least 20% larger than the largest batch of beer you plan to make in order to keep the constituents of the vigorous primary fermentation from spilling out of the fermenter and making a sticky mess. The bucket should be equipped with a lid and gasket, and also have provisions for fitting an airlock. These are available in various sizes from beer and wine hobby suppliers. I recommend a six-gallon bucket.

If you are already making wine, you can use the same primary fermenter you already have and save some money that way.

Don't worry about the smells and tastes of plastic infusing into your beer. This is not a concern because the container is food grade plastic selected for its low diffusion, you cleaned it thoroughly prior to use, and the beer will only be in it for a couple of weeks at most.

Priming Bucket

A priming bucket (also called a bottling bucket) is identical to a primary fermenter except that it has a spigot at the bottom for easily draining the beer into bottles. Priming is a process in which a controlled amount of sugar is added to finished beer before it is bottled. The remaining microscopic yeast in the beer eats the sugar to make carbon dioxide which carbonates the beer.

The reason why a priming bucket is used instead of just bottling straight from the fermenter is because the fermentation process leaves a lot of detritus at the bottom of the fermenter. You want to get your beer away from that, as it contains a lot of yeast cells that will auto-destruct (a process called autolysis) once they have run out of food, and when they autolyze they'll release nasty-tasting compounds into the beer. So the beer is racked from the fermenter into the priming bucket, the sugar is added to the priming bucket, and then the beer is bottled.

Secondary Fermenter

Although a lot of instructions for beer making go straight from the primary fermentation to bottling, I do not believe this gives as high a quality beer as one that has undergone secondary fermentation. Secondary fermentation will allow the beer to clarify naturally as well as develop more complex flavor compounds. Likewise, during secondary fermentation beer will lose the sharp yeast tang that can occur. It's also an excellent opportunity for adding certain flavoring adjuncts.

If you already make wine, you can re-use the wine carboys for beer. If you aren't making wine, you might eventually want a secondary fermenter for your beer anyway, and I recommend glass for this purpose because you can keep the beer in the secondary for as long as a month.

The glass vessels come in various sizes from one gallon up to five gallons. The smaller one-gallon vessels are simply one-gallon jugs, and the larger three-or five-gallon vessels are glass carboys similar to those used on old-fashioned water coolers. You will need a five-gallon carboy.

You will also need to get a special brush for cleaning your carboy because the opening is too small for even the smallest hands and a regular bottle brush is too short and isn't bent for cleaning around the edges.

The fermentation that takes place in the secondary fermenter is long and slow. As the carbon dioxide is evolved more slowly, it is possible for air to be drawn into the vessel, especially if temperatures change. During secondary fermentation, you want to prevent oxygen from coming into contact with the beer, because oxygen adversely affects the quality of the beer by changing the character of some of the evolved organic compounds.

By fitting the hole in the fermenter with a stopper and an airlock, you will allow a protective blanket of carbon dioxide to cover the surface of your beer. You will need rubber stoppers with one hole in them that are sized correctly for your secondary fermenter. The airlock is prepared, put into the hole in the stopper, and then the stopper is placed in the hole at the top of the fermenter.

One thing that people often overlook is a carrying handle. A five-gallon carboy filled with beer is extremely heavy and difficult to handle. The handle that you order can be installed on a carboy and then removed to be used on another, so you only need one. They cost about $10 and are well worth the price.

Airlocks

Airlocks are devices installed on a fermenter that allow gas to escape, but do not allow air to leak back in. They come in a variety of configurations, but all are filled with water or a solution of potassium metabisulfite. The airlock is filled to the level specified on the device, inserted in a one-hole rubber stopper and then attached to the fermenter. You should have at least two of these. The style you choose doesn't matter.

Racking Tube/Auto-siphon

A racking tube is a long two-part tube that is inserted into the beer and pumped to start a siphoning action in order to transfer the beer from one container into another while minimizing contact with oxygen. It has a knob at the bottom that directs the flow of fluid in such a way as to minimize the amount of sediment transferred in the process. You will also need five feet of plastic tubing to go with it. A stop-cock, which is a plastic clip that can be used to stop the flow temporarily, will come in handy.

Always clean your racking tube and plastic tubing before and after use, and run a gallon of sulphite solution through it to sterilize the components. Otherwise, it will accumulate debris attractive to fruit flies that carry vinegar bacteria and you will unwittingly start manufacturing vinegar. The tubing is inexpensive and it is best to replace it after several uses.

Wort Aerator

When wort is boiled, all of the dissolved oxygen is driven out. The problem is that yeast needs oxygen for its initial replicating phase, so the oxygen has to be put back into the wort before the yeast is added. There are a lot of ways to do this, but

I recommend a handy $4 gadget that goes onto the end of the plastic tubing you use to transfer the wort from the brewing pot to the primary fermenter. The gadget sprays the beer all over the place as it goes into the fermenter, giving it ample opportunity to dissolve oxygen. This device is called a Siphon Spray Wort Aerator.

Hydrometer

A hydrometer is used to check the specific gravity of your wort, which gives an indication as to how much sugar is dissolved in it. By measuring the specific gravity before and after fermentation, you can tell how much alcohol the beer has. If you are making wine, you can use the same hydrometer.

Thermometer

In brewing, temperature is everything because so many of the enzymatic processes are dependent upon specific ranges of temperature. I recommend an instant-read digital thermometer, such as the Norpro Model # 5976. You can also use the thermometer in cheese making. In spite of the "instant read" designation of this and similar thermometers, they take a minute or two to give an accurate reading.

pH Meter

You will need to measure and adjust the pH of your mashing and lautering water, and an inexpensive pH meter will make this effortless. Don't forget to get a pH 7 buffering solution for calibrating the meter.

Capper

Because beer is a carbonated beverage, you can't cork it like wine. The caps have to form a good seal and be held securely to withstand the force of compressed gas. Cappers are used to secure the caps on beer bottles. You can buy a bench model or a two handed model. I use a two-handed model that costs $17.

❯❯ Beer uses a special capper and crown caps to contain the compressed gas.

Wine Thief

Even though it is called a wine thief, it is equally useful for beer. A wine thief is a long tube with a special valve on the end that allows you to remove liquids from a container very easily. Clean and sanitize it before and after use. Beer is more susceptible than wine to contamination, so any wort that you remove for testing should be discarded instead of returned to the container.

Beer Bottles

You will need beer bottles. Lots of beer bottles. A beer bottle holds twelve ounces of beer. A gallon of beer is 128 ounces, and five gallons of beer is 640 ounces. That means 53 bottles of beer are needed for a five-gallon batch. In practice, you usually only need 48 bottles because of waste, spillage, and so forth. Even so, that's a lot of bottles.

Once you are making your own beer regularly, you can save and reuse your bottles. But you'll need an initial supply. One good place to get them is from bars. Because a lot of states have deposits, I offer to pay the bar double the deposit for the bottles, and have never been refused. (If you are concerned about potential diseases, beer bottles can be boiled in a boiling water bath canner for thirty minutes after washing.) You can also buy them new, at a cost of $16 per case of 24. As you can see, you'd need two cases for your first batch of beer, which will cost $32. But you'll ultimately start reusing those bottles so the cost for subsequent batches will be zero.

Bottling Bucket

A bottling bucket is a large plastic pail almost identical to the primary fermenter except that it has a spigot at the bottom. The beer is racked into the bottling bucket, the priming sugar is added, and then bottles are filled using a bottling wand attached via a section of PVC tubing.

Bottling Wand

A bottling wand is a spring-loaded device that is used to fill bottles. One end connects to the bottling bucket with a piece of tubing, and the other end is inserted in the bottle to be filled. A bottling wand has a spring-loaded valve at the bottom.

Grain Mill (optional)

You can buy most malted grains already cracked. But if you buy them in whole form or you make your own malted grains, you will need a grain mill. I have a Corona grain mill that I have used for decades for making flour of various sorts, and it does just fine for cracking malted grain when set properly. The cost is about $55, but it isn't a necessity.

Consolidated Equipment List

The following list will make it easy to get everything you will need in the fewest possible shopping trips. I priced this out with a well-known Internet beer and wine hobby shop for $412.50 plus nearly $100 for shipping. At that price for shipping, if you can find the gear locally it is worth the trip. This cost is, of course, a worst-case scenario. I am assuming you don't already have a gas grill that uses propane tanks, a 30-quart pot, or any of the gear for making wine or cheese. If you already have these things, the cost drops dramatically.

Also, most people start brewing with malt extracts and a $100 kit. But pretty soon, if they like making beer, they graduate through various phases to reach the point of making all-grain beers. Along the way, they spend hundreds or even thousands of dollars in trial and error. We're short-cutting that process and taking you straight to the end-game where you save the most money and have the greatest flexibility.

1	30 quart brewpot, aluminum, enamel, or stainless steel
1	Propane burner
1	Filled Propane Cylinder
2	Six-gallon plastic fermenters with sealing plastic lid and grommet
1	Priming/bottling bucket with lid and spigot
1	Five-gallon secondary fermenter, preferably glass
1	Carboy brush for cleaning the secondary fermenter
1	Carboy handle
1	Five-gallon beverage cooler (the best deals are at home improvement stores)
1	Wort chiller
1	Siphon spray wort aerator
1	Hydrometer

1	pH meter
1	Instant-read digital thermometer
1	#6.5 universal rubber stopper with one hole
2	Airlocks
1	Racking tube/autosiphon
1	Bottle filling tube
5 ft	⅜" plastic tubing
1	⅜" hose clamp
1	Wine thief
1	Two-handed capper
48	Beer bottles
100	Bottle caps

In addition to equipment, making beer requires a variety of innocuous but nevertheless important additives. In general, beer yeast are well-adapted to the nutrients in malt and will perform well. However, just like wine, beer is affected by a wide variety of factors and the brewer can exercise some control over these factors with measurements and adjuncts.

Citric, Malic, and Tartaric Acids

You can use any of these to acidify your mash water if the pH is too low, and if you make wine you likely already have these handy. The reason why you would want to acidify the water you use in mashing and sparging is because an acidic water will dissolve fewer tannins from the husks of the grain. They also increase tartness, which may or may not be something you want in a particular beer, which is why other additives are usually used for modifying the pH of the water. Citric acid in particular tends to help with clarifying.

Calcium Sulfate and Magnesium Sulfate

Calcium sulfate and magnesium sulfate are purified food-grade versions of gypsum and Epsom salt respectively. Though calcium sulfate is commonly used in brewing to lower the pH of water, that is not what it is used for in my brewing methods. Rather, calcium sulfate and magnesium sulfate are used for making

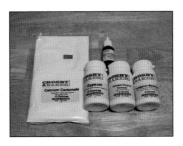

» Adjuncts commonly used in brewing.

very modest additions of calcium and magnesium to water, especially distilled or bottled waters, in which those elements are deficient. Yeast requires certain minimal levels of both calcium and magnesium to function optimally.

Phosphoric Acid

This is the same phosphoric acid used in all manner of soft drinks, bottled iced tea and so forth. It comes as a 10% solution from brewery suppliers. It is inexpensive at $4 for eight ounces. I advocate the use of phosphoric acid to lower the pH of mash and sparging water if needed. This is far easier than the complex calculations of accomplishing this using various positive ions.

Potassium Hydroxide/Wood Lye

You can get food grade potassium hydroxide (also known as reagent grade) from chemical supply stores.[17] A 15g package is more than sufficient. To use it, make a 1N solution by putting 14.03 grams in a 250 ml volumetric flask, and adding distilled water to the fill line. Wear gloves and goggles during this procedure as mixing lye with water can generate heat. Once the solution has been made, it can be stored in a sealed plastic (but not glass) bottle. If the pH of your mash or sparging water needs to be raised (which is unlikely but possible) potassium hydroxide will do a much better job than the potassium carbonate usually used for that purpose, and it will raise the pH without adding hardness.

Pectic Enzyme

This isn't needed when making beer from grains alone, but if you decide to include fruits in your fermenter as an experiment, you'll need to add pectic enzyme in order to avoid the development of hazes in your beer.

---

17 HMS Beagle at www.hms-beagle.com and United Nuclear at www.unitednuclear.com sell reagent grade potassium hydroxide in 15g and 2 oz sizes respectively for less than $5.

Amylase Enzyme

Malts used in making beer already contain amylase. The only reason you would add this is if you decide to add grains to your mash that have not been malted such as oats. In general, light malts have enough residual enzymes to convert the starches of a reasonable amount of adjunct grains. But if you are using relatively little malt or the malts you are using are predominantly dark, then the natural enzymatic activity will be insufficient and a little boost will be needed. Add one teaspoon to the mash and mash as usual.

Clarifying Agents

Because wine spends so long in secondary fermentation, a time ranging from months to years, wine has plenty of time to clarify on its own without the need for filtering or added clarifying agents. Beer, however, seldom has enough biological stability to spend more then three weeks—four at the most—in combined primary and secondary fermentation. So clarifying agents are often used right in the brew pot or at least added during secondary fermentation in order to help clear it of suspended yeast particles, proteins, and solids. Though in the near term suspended particles just give a bad appearance, once beer has sat around for a while they will break down and impart off-flavors. So you want to clarify your beer.

A bewildering array of clarifying agents (also called finings) are available, but I recommend those that are easiest to use with the widest range of effectiveness: Irish moss, gelatin, and Polyclar (PVPP).

Irish moss is a brown seaweed also known as carrageen—a substance you have likely already consumed in ice cream, dressings, and so forth. It is inexpensive, and very effective at coagulating proteins. Add a quarter teaspoon of Irish moss to the brew pot fifteen minutes before the end of boiling, at the same time you would insert the wort chiller.

With a week to go in the secondary fermenter before bottling time, you should add a mixture of gelatin and Polyclar (PVPP) to the secondary fermenter.

Gelatin will help prevent protein hazes and will condense any dead yeast that is floating around. To use gelatin, use your scale to measure out one gram of fining gelatin (from a wine/beer making store), and mix that with two tablespoons of cold water in a clean coffee cup. Separately, put seven tablespoons of water in a glass measuring cup, and heat on high in the microwave for one minute. Add the hot water to the dissolved gelatin in the coffee cup, mixing thoroughly. Allow this to cool down to a temperature of 80 degrees, and then gently stir the whole amount into five gallons of beer.

You should add Polyclar (PVPP) at the same time. These are tiny statically-charged beads of harmless plastic that will attract any remaining particles not already attracted by the Irish moss or gelatin. Mix ¼ teaspoon of Polyclar into the gelatin before adding it.

Yeast Energizer

Yeast energizer supplies crucial nutrients for yeast that allow it to reproduce and do a good job of converting sugar to alcohol. Yeast energizer usually contains food grade ammonium phosphate, magnesium sulfate, yeast hulls to supply lipids, and the entire vitamin B complex; of which thiamine (vitamin B_1) is the most important. Grain contains a lot of nutrients, so this isn't usually needed as it would be with country wines. But it also doesn't hurt to add it anyway.

Sulfite

Because beer wort is effectively sterilized by boiling, there's no point in adding sulfite directly to the wort as is done in wine. Even so, sanitation is even more important in beer than it is in wine due to both its nutrient density and lower alcohol content, and sulfite is well employed in sterilizing equipment.

For brewing you only need one form of sulfite: powdered potassium metabisulfite. In powdered form it is used to make sterilizing solutions for sterilizing equipment.

Yeast

The only thing with more yeast varieties available than wine is beer. The array is staggering. Yeasts for making beer fall into two primary categories: ale yeasts that ferment at the top of the fermenter and prefer temperatures above 60 degrees, and lager yeasts that ferment at the bottom of the fermenter and prefer temperatures below 60 degrees. They are available in three forms: dried packets for direct pitching, liquid tubes requiring amplification before pitching, and liquid foils for direct pitching.

Whereas with wine, dried yeast packets are inexpensive and available in nearly inexhaustible variety, but with beer there aren't many varieties available.

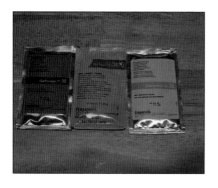

Luckily, those that are available are of high quality. Fermentis makes seven very flexible dry yeasts.

Wyeast direct pitching yeasts come as an inner sealed packet of yeast in an outer sealed packet of nutrient. They are activated by breaking the inner packet according to package directions, and then setting the packet aside for three hours or more while the yeast multiplies. Be sure to sanitize the package with sulfite solution before opening and pitching.

Other than the temperature requirements, making lagers is not appreciably different than making ales, but because most homes aren't maintained at temperatures suitable for lager, I am going to specify three good general purpose ale yeasts: one in dried form and two in foil packet form.

Fermentis SafBrew S-33 Dry Ale Yeast

This is a very good all-around yeast that develops excellent flavors. S-33 performs best when reconstituted prior to pitching. To reconstitute, add the contents of the packet to four ounces of sterile (boiled and cooled) water at a temperature of between 74 and 86 degrees. Keep in a warm place, and stir periodically with a sterile spoon until a slurry has been formed. After an hour from when the yeast and water were first mixed, gently distribute into the wort in the primary fermenter.

Wyeast 1056 American Ale

This is a great yeast for making beer that is unmistakeably beer. That is, it produces very few fruity or ester aromas, and allows the flavors of the underlying malt and hops to predominate. This yeast is very flexible, and can be used for making a variety of ales, porters, stouts, and even braggot or barleywine.

Wyeast 1084 Irish Ale

This yeast is almost the opposite of 1056. Especially when fermentation is conducted between 64 and 70 degrees, it produces a lot of esters and fruit notes. It is especially well-suited to making dark and robust ales and stouts.

Consolidated Ingredient List

The following ingredient list will allow you to make many successful gallons of beer. As your experience expands, you may wish to adopt different materials and techniques; but most home brewers find that this list is more than sufficient for their needs. The total cost of all these supplies will be well under $60.

4 oz	Citric acid
½ oz	Pectic enzyme liquid
4 oz	Amylase enzyme powder
2 oz	Yeast energizer
1 oz	Irish Moss
1 oz	Brewer's Gelatin
½ oz	Polyclar
2 oz	Calcium Sulfate
2 oz	Magnesium Sulfate
4 oz	10% Phosphoric Acid
15g–2 oz	Reagent grade Potassium Hydroxide flakes
4 oz	Powdered potassium metabisulfite
1 pkt	Fermentis SafBrew S-33 ale yeast
1 pkt	Wyeast 1056 ale yeast
1 pkt	Wyeast 1084 ale yeast

8

The Science of Beer

The science of beer is only slightly more complex than the science of wine. Unlike wine in which the sugars exist naturally, the sugars used to ferment beer must be produced by enzymatic action. Once the sugars have been produced, the fermentation itself works the same.

Enzymes

Enzymes are biological catalysts. A catalyst is a substance that assists in bringing about a change in some other substance without being changed itself. A common example of catalytic action is the catalytic converter mandated in modern cars. Once it heats up, the catalysts it contains burn any unburnt fuel leaving the engine so that it won't contribute to pollution and fog. Enzymes within our bodies work much the same, assisting with nearly every body and cellular action. It is the enzymes in malt that are used to create fermentable sugars from starch.

Making Malt

The core of making beer is malted grain. When grain begins to sprout, it produces enzymes that serve to convert the stored starches in the kernel into readily metabolized sugars for the emerging plant. If the grain continues to sprout, those sugars are used by the plant and the enzymes are destroyed as they have outlived their purpose. Malting arrests the germination of the seed at precisely the right time for the maximum presence of enzymes.

Three grains are well-suited to making malt: barley, wheat, and corn. Though all grains make the appropriate enzymes to some degree, it is these three whose conversion is most efficient. Barley is the most used because it not only makes enough enzymes to convert its own starches, but it also makes enough to convert the starches in a modest amount of unmalted adjunct grain.

The grains are induced to sprout through a cyclical soaking regimen: soaked in water for eight hours, then removed from the water for eight hours. During the time they are out of the water, they are spread evenly on a flat surface and protected from drying either by enclosure or occasional light misting. This process is repeated until the root emerges and the acrospire—the baby leaf that is inside the grain kernel—is at a length of between 75% and 100% that of the grain. Once the root has reached the length of the grain, you just open a sample grain and look inside to check on the length of the acrospire. The acrospire is white because it has not yet emerged into light and may be hard to see. This whole process will take around six days for barley and about three days for wheat and corn.

Once the acrospire is long enough to indicate enzymes are at their peak, the germination is halted by kilning, in which the germinating grains are heated.

Wheat and corn are heated at a temperature of 110 to 125 degrees until they weigh the same as they did before germination. Barley is heated at that temperature for a day, but then the temperature is gradually raised to 140 to 160 degrees until such time as the barley weighs the same as it did before germination. This process gives you pale malt, the type of malt with the greatest enzymatic activity. Other malts gain their darker colors from roasting and impart other flavors and textures, but generally have a lesser degree of enzymatic activity because the higher temperatures denature the enzymes.

Using this information you can, of course, make your own malt from whole grains—the ultimate self-sufficiency in beer making, especially if you grow the grains yourself! But the primary reason I covered this is so malts don't seem mysterious, inaccessible, or something that you have no choice but to depend upon others to supply.

Enzymes and Malt

Malt contains a number of enzymes of interest, each of which has different effects, different optimum temperatures and work best at different pH values.

β-glucanases

β-glucanases are responsible for tearing down the cell walls of the grain. Tearing down the cell walls makes what is inside the cells available. They work best at a temperature of 95 to 113 degrees. In general, these enzymes have already done all the work they need to do during the first stage of the kilning process, so there is no need to mash at this temperature. If you are using homemade malt and you mash at this temperature range for fifteen minutes, this is called a beta glucans rest.

Proteases

Proteases are responsible for breaking the bonds between amino acids that hold protein chains together. This turns the proteins in grain into amino acids that can be used by the yeasts for their own reproduction. The most active range for protease enzymes is from 113 to 131 degrees. Normally, the temperature of malt is raised gradually, so it passes through this range slowly enough while it still has enough water content, which means the job of the proteases has been completed in the malt. In this case there is no need to mash at this higher temperature. However, if you are using homemade malt, mashing for twenty minutes between 122 and 125 degrees would be a good idea before increasing the temperature for the rest of the mashing process. This aspect of mashing is called a protein rest.

α-amylases

Alpha amylases are present in human saliva in the form of ptyalin and also in pancreatic juice, in many fungi and in seeds such as grains that use starch as an energy reserve. In all of these cases, the purpose of alpha amylase is to liquify starch by breaking the long chains into shorter chains. Though α-amylases are active from 140 to 167 degrees, they work most quickly at 158 degrees.

β-amylases

Beta amylases are present in various plants, bacteria, and fungi. They are responsible for breaking up the starches in ripening fruit into sugars. They are also present in grains used for malt, where they break down the liquified starch into maltose (sugar) molecules. They work best at 140 to 145 degrees, but are active from 140 to 150 degrees. Beta amylases are denatured within twenty minutes at temperatures exceeding 155 degrees.

Mashing

Because proteases and β-glucanases have already done their respective jobs during the malting process, most mashing is optimized for the action of α and β amylases. The fact that they work best at different temperatures allows us to control the alcohol content and mouth-feel of the beer by the temperature at which it is mashed.

In practice, then, the mashing temperature is somewhere between 149 and 159 degrees, and the temperature of the mash profoundly affects the character of the beer. Toward 149 degrees, you end up with a thinner mouth-feel to the beer and higher alcohol content because of the greater conversion to fermentable sugars, and toward 159 degrees the short chain starches give a thicker mouth-feel to the beer, but the alcohol content is lower.

For most commercially available malted grains, what is called a single-infusion mash is adequate and typical. In a single-infusion mash, the grains are added to the mash tun and a single temperature between 149 and 159 degrees is maintained depending upon the type of beer desired. If a beer with a lot of mouth feel is desired but this doesn't give enough fermentable sugar, an adequate amount of adjunct sugar is mixed with the wort when it is boiled.

For a single infusion mash, the cracked malt is put in the mash tun (in my case a $20 beverage cooler from a home improvement store) and water ten degrees warmer than the desired temperature is added. Simply heat up the water in a big pot on the stove, measuring the temperature until it is ten degrees warmer, and then pour it into the mash tun with the grain. (Add water to the grain rather than grain to the water as the enzymes work better this way.)

A quart to a quart and a half of water is needed for every pound of malt. You need to heat the water ten degrees hotter (seven if using a quart and a half of water per pound of malt) because the malt at room temperature will lower the temperature of the added water. After adding the water and stirring, check the temperature and adjust by careful additions of no more than a quart of hot or cold water.

You can "have your cake and eat it too" by using what is called a step mash. If you are having trouble with protein hazes and head retention in your beer, a step mash may be called for. A complete step mash is started in a brewing pot for easily increasing the temperature, and ends in the mash tun for the final conversion when the highest mashing temperature has been reached.

A typical step mash spends 10-30 minutes at 95 degrees for β-glucanase activity (even in step mashes this is often skipped), 10-60 minutes at 122 degrees for protease activity (called a protein rest), 20–30 minutes at 148-150 degrees to maximize fermentable sugars with β-amylases, and then the temperature of the mash is raised to 155–158 degrees for the mouth-feel generated by α-amylases and dumped into the insulated mash tun; hold the final temperature for 30-60 minutes. The final mashing time is over when all starch has been converted to either sugar or shorter starch chains. This is checked using an iodine test.

Iodine Test

Who would have thought that Mrs. Breckinridge's seventh grade science class would be so useful? In that class, along with dissecting frogs, I also learned something eminently useful for making my own beer: the starch reaction.

To perform the iodine test, put a small one teaspoon sample of mash wort on a clean, white saucer and a drop of tincture of iodine on a separate part of the plate. Gently tip the saucer so that the mash makes contact with the iodine and observe the color. If it is black, blue, or deep purple, there are still unconverted long-chain starches. If it is red, brown, or reddish brown, all the long-chain starches have broken down into sugars and short chain starches. If the mash is completely converted to sugars, the iodine remains yellow. As long as the test shows that all long-chain starches have been broken down, it is fine to proceed to sparging.

Water

Water constitutes between 90% and 95% of beer by weight, so it is important to use high-quality water when making beer. Not to disparage the efforts of good people who have written hundreds of articles on managing water for making beer, but 95% of the time you can brew fantastic beer with water from your faucet or using bottled spring (not distilled) water with only minimal management.

Chlorine and Chloramines

The water used for brewing must be free of chlorine and chloramines. These substances are used as disinfectants in municipal water supplies, but in beer they combine with natural phenolic compounds to create nasty-tasting beer. Chlorine can be removed with a carbon filter, and chloramines can be removed by adding ¼ of a Campden tablet per 5 gallons of water and allowing to stand for an hour.

Angst-Free pH Management

pH is important because the enzymes in mashing work best within certain pH ranges. Water that is too alkaline (i.e. has too high a pH) will absorb too many tannins from the grain while mashing and sparging. In general, if your water is good for drinking and isn't very hard, its chemistry is such that when it is mixed in the mash tun, the natural phosphates in the mash will combine with cations in the water to create a pH somewhere in the ideal range of 5.2 to 5.8.

Though it is possible to adjust the mineral content such that based upon certain grains, toasting level, adjuncts, starting water chemistry, and so forth you will hit a certain pH, the task is complex and subject to error. If a mistake is made, the beer ends up tasting like it was brewed from mineral water.

Instead, I recommend using your pH meter to test the pH of the mash, and to adjust the pH right in the mash tun based upon that reading if needed. You want a reading between 5.2 and 5.8. Most levels will be fine, but if it is too high, you can lower it through addition of dilute phosphoric acid. If it is too low, you can raise it with the addition of potassium hydroxide solution. Both of these were described (including how to make the potassium hydroxide solution) in the previous chapter.

Keep in mind that on the first test it will be impossible to know how much of these to add to change the pH, because many compounds in mash serve as buffers. A buffer is a compound that takes up some of the acid or base that you are adding in such a way as to avoid a significant change in pH. Once the buffer is used up, the pH changes fast. Add 0.5 ml (measured using a syringe without the needle) of either phosphoric acid (to lower the pH) or potassium hydroxide (to raise the pH), mix into the mash, wait a minute, and test the pH again with the meter. Repeat until the pH is within the range of 5.2 to 5.8.

Sparging water doesn't have the same opportunity for contact with the mash, and so is more likely to need pH adjustment. In fact, most professional breweries pre-treat their sparging water to a pH of 5.5, and I recommend that you do the same. A pH of 5.5 will ensure that the sparging water won't dissolve an inordinate amount of tannin from the grain husk. To adjust the pH of sparging water, do it right in the kettle once it has reached sparging temperature. Use the same technique as for mash water, making 0.5ml additions of either phosphoric acid or potassium hydroxide and checking with the pH meter after each addition. Because plain water doesn't buffer as well as mash water, as you approach within half a point of the desired pH, decrease the amount of your additions to 0.2ml.

Minerals

If your water is good for drinking and if it already makes good beer, the old adage "if it isn't broken don't fix it" applies. You can brew beer far superior to what you can buy without ever messing with your water. Where messing with your water makes a difference is on the high end of beer competitions. Under those circumstances, just a tiny difference in ion concentration in the sparging water can make a detectable difference in a competition. But for brewing a good solid beer that will beat the store any day, you seldom need to bother with this.

All of the foregoing notwithstanding, there's just something about brewing beer that makes brewers love tinkering with the mineral content of the brewing water, and if you can't resist the urge, here are some guidelines.

First things first: Before you can tinker, you need to know what you have. For this, you need a water test. If you are using bottled spring water, you can get that information from the manufacturer's website on the Internet. If you are using municipal water, you can sometimes get that information from your water provider. If they won't provide it or if you are using a private well, you will need to send out your water for testing.

There are many water testing laboratories and I don't own stock in any of them, but the one most often used by brewers is Ward Labs at www.wardlab.com. The test you want is "W-6 Household Mineral Test" for $16.50. I am going to cover each parameter tested and give some guidelines and information that will be useful regarding each one.

Sodium: (Na+) Just as it does with other foods, sodium rounds out the flavor in beer but tastes salty in excess. The amount of sodium in your water can go down to zero and as high as 150 mg/L without adverse effect, but you'll definitely notice a salty taste with concentrations exceeding 200 mg/L.

Calcium: (Ca2+) Yeast and mash enzymes both require a certain minimum level of calcium to function efficiently. In general, it is accepted that the minimum level of calcium is 50mg/L and the maximum level is 150mg/L.

Magnesium: (Mg2+) As with calcium, the yeast and mash enzymes require a minimal level of magnesium in order to function efficiently. The minimum level of magnesium is 10mg/L, and the maximum level is 30mg/L. Levels greater than 50mg/L will make beer taste sour.

Potassium: (K+) The level of potassium should be less than 10mg/L. Otherwise, the high level of potassium will inhibit yeast.

Chloride: (Cl-) Chloride isn't the same thing as elemental chlorine. It is an ion that is usually paired with sodium, potassium, calcium, or another element. There's no requirement for chloride in brewing water, but if levels exceed 250 mg/L you can end up with salty-tasting beer.

Sulfate: ($SO_4$2-) Sulfates contribute to what is called "permanent hardness" in water. Permanent hardness refers to hardness that can't be removed through boiling, precipitation, or other means. At levels up to 150 mg/L the beers will come out as expected, but at levels greater than 150mg/L, your beer will come out surprisingly bitter.

Carbonate/Bicarbonate: (HCO_3-) The form in which carbonates/bicarbonates exist in water is dependent on the pH of the water. When the pH exceeds 8.4, carbonates predominate. With a pH of of less than 8.4, bicarbonates predominate. The level of bicarbonates has an effect on the type of beer you can brew. At levels below 50 mg/L you can brew pale beers, at levels greater than 150 mg/L you can brew dark beers, and from 50mg/L to 150mg/L you can brew amber beers. You can't brew effectively at levels greater than 250mg/L, but luckily by boiling the water for 30 minutes and carefully decanting you'll leave all but 50mg/L of the bicarbonate behind in the form of insoluble carbonates.

When the levels of a particular element are too high, you can purchase bottled water to dilute it sufficiently to bring that element within an acceptable brewing range. That part is easy and it's pretty rare to get into trouble that way.

When you are trying to tweak the levels of an element higher, don't forget that these elements come paired (for example as magnesium and sulfate in mag-

nesium sulfate) so that by adding one you are automatically including the other. The following table of common brew water additives tells you how many mg/L of each ion you get by adding a gram to five gallons of water.

Mineral Salt	Ca	Mg	Na	Sulfate	Chloride	Carbonates
Calcium Chloride	14	-	-	-	25	-
Calcium Sulfate	12			30		
Calcium Carbonate	26			32		
Magnesium Sulfate		5		20		
Sodium Bicarbonate			15			38
Sodium Chloride			21		32	

Mineral additives and mg/L of Added Elements per Gram Added to a Five-Gallon Batch

As an example, pretend I am using Monadnock Mountain Spring Water, and I want to make an amber ale. The analysis from the company's website shows a calcium content of 3.4mg/L, hardness of less than 10 mg/L, chloride of 2.5mg/L, and essentially nothing else. I'd like my water to have a bit of character, with elements that at least exceed the minimum thresholds plus be in the range of carbonates needed for an amber beer.

I'm going to add three grams of magnesium sulfate for 15 ppm of magnesium and 60 ppm of sulfate, three grams of sodium bicarbonate for 45 ppm of sodium and 114 ppm of carbonates, and four grams of calcium chloride for 56 ppm of calcium and 100 ppm of chloride.

Hop Chemistry

When I was a kid, I asked my dad what was used to make beer. He said "hops." As a boy who enjoyed catching all manner of grasshoppers and crickets, I had an image in my mind that a "hop" was likely to be a tiny grass-eating insect of some

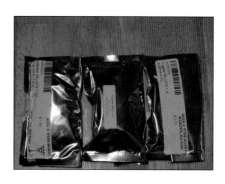

❯❯ Some hop pellets with alpha-acid content noted on the label.

sort. A bunny rabbit didn't sound plausible. Just imagine my surprise when I learned that hops are actually the flower of a climbing vine.

Hops are added to beer for two purposes: to provide a bitterness that balances the sweetness of malt and as a preservative.

Hops contain a number of important compounds. Referenced most frequently are the alpha acids: cohumulone and adhumulone. These acids are responsible for the ability of hops to impart bitterness, and hops are usually sold with a notation as to the percentage of alpha acids they contain. In their natural form, alpha acids are not water soluble, but when hops are boiled long enough the alpha acids isomerize. The change in structure makes them water soluble, and the isomerized acids have a bacteriostatic effect on gram-positive[18] bacteria such as lactic bacteria.

Beta acids in hops include adlupulone, colupulone, and lupulone. Unlike the alpha acids, boiling the hops doesn't cause the beta acids to isomerize into a water soluble form. These acids form an important part of the aroma of bitterness, though, so hops being used for their beta acid content are added late in the boil so the beta acids aren't boiled away.

Hops also contain a number of essential oils that contribute non-bitter flavor and aroma, including caryophyllene, farnesene, humulene, and myrcene. These essential oils aren't found only in hops. They also contribute to the aromas of plants as diverse as bay, ylang-ylang, thyme, coriander, cannabis, rosemary, cloves, black pepper, and gardenia, among others. Though they are usually added late in the boil or even added to the primary fermenter without boiling (a process known as dry hopping), the differences in boiling point and solubility between the various essential oils can create differences in the flavor of beer with even slight differences in technique. Each variety of hop used for aroma also starts out with different relative proportions of these components.

If all you have ever tried for beer is commercial varieties, then you might well be skeptical of people claiming to taste citrus, flower, or other components in

---

18 The term "gram-positive" refers to a broad categorization of bacteria as either gram-positive or gram-negative depending upon whether their cell walls are stained dark blue or purple by gram stain.

home-crafted beers, but the fact is that hops have the ability to impart a wide array of flavors and scents. In the next chapter we'll examine different hop varieties and their uses.

Fermentation

The term "dextrose" is used for the sugar created in malting. It just so happens that dextrose is a synonym for glucose, so the fundamental chemical equation for turning sugar into alcohol in beer is identical to the equation for wine:

$$\text{glucose} \rightarrow \text{ethyl alcohol} + \text{carbon dioxide} + \text{energy}$$
$$C_6H_{12}O_6 \rightarrow 2(CH_3CH_2OH) + 2(CO_2) + \text{energy}$$

Nearly all of the sugars in wine are fermentable, whereas many created in malting and mashing are not. These non-fermentable sugars provide body and sweetness to beer.

Specific Gravity

The specific gravity of beer is measured using the same methods as for wine, but the meaning of the measurements is very different. In wine musts, sugar is so overwhelmingly present compared to other dissolved solids that the specific gravity translates directly into sugar concentration with minimal error. As a result, when a wine is completely fermented, it will end up with a specific gravity of less than one. This low specific gravity is a result of the fact that the alcohol in the wine has a lower specific gravity than water.

Beer wort is a more complex mixture; it contains a variety of both fermentable and non-fermentable sugars, amino acids, short-chain starches, and other components that increase its specific gravity. As a result, measurements of starting specific gravity alone give no indication of the amount of fermentable sugar in the wort, and therefore no indication of alcohol content. Furthermore, the specific gravity of beer that has completed fermentation will be greater than one, because even though its fermentable sugar has been converted to alcohol, the short-chain starches and non-fermentable sugars remain.

Specific gravity readings are taken while lautering so you know when lautering has been completed. They are also taken when the wort is first placed in the primary fermenter (this measurement is the original gravity or OG) and when the beer is bottled (this is called the final gravity or FG). The difference between the OG and FG is called attenuation.

This is expressed as a percentage using the following formula:

$$\text{attenuation} = 100\% * (OG{-}FG)/(OG{-}1)$$

Attenuation is the percentage of potential sugar that has been converted to alcohol. It gives a good estimate of alcohol content, but is primarily a measure of the entire brewing process, as everything from the malting to the mashing to the yeast variety will have an effect on attenuation. A higher degree of attenuation indicates a beer that has more alcohol and less sweetness, whereas a lower degree of attenuation indicates a beer with lower alcohol but greater sweetness.

You can tell approximately how much alcohol your beer has by using a hydrometer table to subtract the potential alcohol reading that corresponds to your final gravity from the potential alcohol reading corresponding to your original gravity.

Another way is to use the following formula:

$$\text{Alcohol Percent by Volume} = 117 * (\,(OG{-}1){-}(FG{-}1)\,)$$

For example, if the starting gravity of your beer is 1.068 and the final gravity is 1.017, the alcohol content is: $117 * (\,(1.068{-}1){-}(1.017{-}1)\,) = 5.97\%$

Conditioning

Beer can be bottled after the primary fermentation if the fermentation is allowed to proceed for a couple of days after its obvious completion. It is racked into the priming bucket, the priming sugar is added, and the beer is bottled.

Beer handled in this way is fine for drinking so long as it is consumed within a couple of weeks after it has developed adequate carbonation. It doesn't keep very well because a lot of visible particulates remain in the beer, and those particulates start to autolyze and degrade, imparting off-flavors to the beer.

Likewise, beer handled in this fashion is called green because it has not had an opportunity for the flavors to meld and combine. Aging in the bottle helps a bit, but there is a race between aging and degradation. To solve these problems, a number of conditioning methods have been developed to give beer better melded flavors, fewer particulates, and longer shelf life.

Clarification

Clarification is the process of removing particulate matter and hazes from your beer. A secondary fermentation, in which the beer is racked from the primary fermenter into a glass carboy as a secondary fermenter, can go a long way towards allowing solids to settle.

Most often, the steps to clarify beer start right in the boiling kettle. Irish moss is routinely added fifteen minutes before the end of boiling at the same time you would insert the wort chiller. Irish Moss is a brown seaweed also known as carrageen—a substance you have likely already consumed in ice cream, dressings, and so forth. It is inexpensive, and very effective at coagulating proteins.

For beers that won't be subjected to secondary fermentation, Irish moss is usually sufficient. As the primary fermentation dies down, the moss helps coagulate the proteins, and they fall to the bottom of the fermenter. The material at the bottom of the fermenter is called trubb. When you rack the beer from the primary fermenter into either a secondary or the priming bucket, the trubb is left behind.

For beers that will be subjected to secondary fermentation, I recommend a combination of gelatin and PVPP. Add a mixture of gelatin and Polyclar (PVPP) to the secondary fermenter, one week before bottling.

Gelatin prevents protein hazes and condenses any dead yeast cells in the beer. To use gelatin, measure out one gram of fining gelatin, and mix that with two tablespoons of cold water in a clean coffee cup. Separately, put seven tablespoons of water in a glass measuring cup, and heat on high in the microwave for one minute. Add the hot water to the dissolved gelatin in the coffee cup, mixing thoroughly. Allow this to cool down to a temperature of 80 degrees, and then gently stir the whole amount into five gallons of beer.

I recommend adding Polyclar (PVPP) at the same time. These are tiny statically charged beads of harmless plastic that will attract any remaining particles not already attracted by the Irish moss or gelatin. Mix ¼ teaspoon of Polyclar into the gelatin before adding it.

Other clarifying agents are available; use them according to the manufacturer's directions.

Secondary Fermentation and Lagering

Just as with wine, the secondary fermentation is not primarily for the conversion of sugars to alcohol, because 99% of that occurs in the primary fermenter. Secondary fermentation isn't really a fermentation process at all—it is a conditioning process. Secondary fermentation allows for yeast and particulates to settle out and for the combination and melding of various flavoring compounds. Overall it can provide greater biological stability of the beer. For ales, secondary fermentation takes place at room temperature and for a period that is twice as long as the primary fermentation, not exceeding sixteen days. So if your primary fermentation was four days, you can carry out a secondary fermentation for anywhere between eight and sixteen days.

Beer is light-sensitive. Exposure to light will literally give it flavors similar to skunk. Most secondary fermenters are glass, so make sure that glass fermenters are kept away from sunlight. You might even consider covering the secondary fermenter with a blanket.

Lager-type beers are made similarly to ales at first, but they are fermented using a form of yeast that is adapted to fermenting at the bottom of the container and that requires lower temperatures. The primary fermentation is undertaken at under 60 degrees, and the secondary fermentation, in this case known as lagering, takes place at temperatures as low as 35 degrees.

The temperature of the primary lager fermentation is dictated by the directions pertaining to the particular yeast strain you are using. The temperature of the secondary fermentation known as lagering is ten degrees lower, down to a minimum of 35 degrees.

Beer is a fertile medium for accidental inoculation with undesired organisms, and because of its low alcohol level, it is nowhere near as biologically stable as wine. Furthermore, chemical reactions including those associated with aging take place at faster rates as temperature rises. So the length of time you can lager is dictated by the temperature at which lagering is done. At 35 degrees, you can lager as long as eight weeks. At 40 degrees, you can lager for six weeks. At 45 degrees, you can lager for no longer than four weeks.

Maintaining lagering temperatures is difficult in the typical home. Brewers who choose this method of conditioning, which produces the smoothest beer flavors, typically employ a dedicated refrigerator with a supplemental thermostat that will allow it to operate at temperatures suitable for lagering.

Though lagering is usually employed with beers made with bottom-fermenting lager yeast, you can also employ the technique with ales made with top-fermenting yeast. Because the ale yeast won't work at temperatures that low, its primary purpose would be to allow for clarification and the smoothing out of flavors.

Bottling and Carbonation

Beer should be bottled in thick, high-quality bottles that will withstand the pressure of carbonation. The bottles should also have a dark color to block out light and preserve the quality of the beer. Once beer is bottled, if it is well-made initially, it will be suitable for drinking after two or three weeks, and its quality will improve in the bottle for up to three or four months, after which quality will start to decline.

Beer to be bottled is racked into what is called a priming bucket or bottling bucket. This is nothing more than a plastic pail like the primary fermenter that has a spigot on the bottom. The reason why a bottling bucket is also called a priming bucket is because it is used for adding priming sugar for purposes of carbonation. Priming sugar gives the residual yeast in the beer a small snack, and once the beer has been capped that small amount of sugar is consumed to generate carbonation under pressure.

To carbonate five gallons of beer, either five ounces (¾ cup) of corn sugar[19] or one cup of malt extract is added to two cups of water in a small sauce pan and boiled to assure sterility. After boiling, the pan is covered with a lid to avoid contamination and placed in cold water in the sink (Don't let any get in the pan!) to cool it off to below 80 degrees before it is added to the beer in the priming bucket and carefully mixed in with little splashing.

Five ounces of corn sugar is sufficient to provide carbonation at a level identical to that employed by most American lagers—about 2.65 volumes of CO_2 per volume of beer. You can use more priming sugar to get more carbonation, and less priming sugar to get less carbonation. Some styles of beer, such as Scottish ales, only use one volume of CO_2 per volume of beer and others, such as German wheat beers, can use as much as 4.3 volumes of CO_2 per volume of beer. Standard beer bottles can withstand three volumes of CO_2 at standard room temperatures. Beyond that and you'll want to use heavier bottles.

The following equation gives the amount of corn sugar added to a five-gallon batch to achieve any level of carbonation of one volume or greater at 75 degrees.

Ounces of Corn Sugar for 5 gallons = (Volumes of CO_2 × 2.65) – 2

After the priming sugar has been mixed with the beer in the priming bucket, the beer is transferred into bottles using a bottling wand. A bottling wand has a spring-loaded valve at the bottom. You insert it in the bottle, pressing it against the bottom until the bottle is full and then withdraw the wand. This puts the correct amount of beer in the bottle, and leaves the correct amount of airspace. After filling, put the caps loosely on the bottles for fifteen minutes before capping them. This gives the carbon dioxide in the beer time to displace some of the oxygen from the atmosphere before capping, and thereby allows the beer to keep better. Please see Chapter 10 for step-by-step instructions on priming and bottling.

---

19 ½ cup of honey may be substituted.

9

Beer Ingredients and Recipes

Though I spent the last chapter describing beer as a science, it is really an art. The science just tells you how and why, but art is what will inform your choices of how to combine these things.

The potential ingredients, proportions, and combinations in beer are effectively infinite, but the basics can be laid out so that you can mix and match. There are already books dedicated to duplicating existing styles of beer. Rather than cover that material, I am instead going to cover the ingredients and how they are combined in such a way that you can make your own styles of beer.

Malt

Malt comes in two basic forms: base malt and specialty malt. Base malts are often light-colored because they haven't been roasted enough to disrupt their enzymes, and they serve as the base for recipes. Specialty malts have been roasted, smoked, or otherwise treated so that they convey different flavors and colors. Typically, their enzymatic power has been reduced or destroyed depending upon how they have been treated.

Some, such as crystal malts, have already converted their starches to sugars before kilning to kill the enzymes, so they provide fermentable sugar to the wort when mashed with the base malt. The darker their roast, the more their sugars add sweetness because they become increasingly unfermentable with heat exposure. Other specialty malts are practically unconverted and serve primarily for flavor and color, though some tiny amount of their unconverted starches may be converted in the mash if they are combined with a base malt of sufficient reserve enzymatic activity.

The color of malt is given in "degrees Lovibond." The pale lagers and pilsners commonly consumed are 2 or 3° Lovibond, whereas the pitch black imperial stouts are 70° Lovibond. Malts can have colors as dark as 500 degrees Lovibond. They are added to the wort in very small quantities and the dilution makes the color lighter. Finished beer has less color than the wort by about 30%.

Base Malt

In the prior chapter I discussed how malt is made so that you can make your own by sprouting grains. What I described is the procedure for making base malt. A base malt has not been heated to such a degree as to destroy or diminish its enzymes. As a result, base malt is the core ingredient from which beer is made. Not only does it have enough enzymatic activity (also called diastatic power) to convert all of its own starches to sugar, it has enough left over to convert the starches in adjunct grains.

Diastatic power is given in "degrees Lintner." For grains to be effectively utilized in the mash, the combined diastatic power of all grains should be 40° Lintner or greater.

Base malts are not roasted, so they give rise to light-colored beer unless mixed with something darker. A downside of base malts is that a compound called dimethyl sulfide (DMS) is generated when they are mashed. This compound, if

> ❯❯ Some malts used in our example recipe.

it remains in the wort, will give rise to various undesired flavors including that of corn, onions, and cabbage depending upon concentration. Luckily, DMS is volatile and when the wort is boiled, it is driven off. It is important that the wort be cooled quickly using a wort chiller, not just for sanitation reasons, but because cooling it quickly prevents the wort from reabsorbing the DMS.

There are dozens of base malts, but there are four styles that lend themselves well to beginning efforts. North American two-row is made from two-row barley grown in the U.S. or Canada. Two-row barley has a lot of protein and a typical diastatic power of 140° Lintner. When using this base malt, you can use adjunct non-malted grains of up to ⅓ of the total grain. Common varieties include Harrington and Metcalf, and it is common for these to be sold blended. This is one of the least expensive malts, and you can make five gallons of beer from it for about $10. It is very pale, less than two degrees Lovibond, so it is common for brewers to add some specialty malt to darken the beer. Because of its high protein content, it should be step-mashed using a protein rest.

North American six row malt has an extremely high enzymatic power of 160° Lintner, allowing for as much as half the total grain in a recipe to be unmalted adjunct. This makes it very popular with mass-market breweries who like to include cheaper rice and corn in their mash to save costs. Especially if adjunct grains are used, a protein rest will be needed to convert the proteins in the grains to free-form amino acids required by the yeast. This base malt is pale, at two degrees Lovibond or less.

British pale ale malt has a color of 4° Lovibond, is well-converted, and with a diastatic power of only 40-65° Lintner it has very little enzymatic potential beyond what is needed for its own conversion in mashing. However, its low protein content and high conversion make it one of the simplest base malts to use as it lends itself well to single infusion mashing. It doesn't work well with unmalted adjunct grains, but it is fine with the addition of specialty malts for added flavor and color.

German and Belgian Pilsner malt is very pale at two degrees Lovibond and has a diastatic power of 110° Lintner. Though it is used predominantly in beer styles of the same name, its high diastatic power lends it to use in any number of recipes.

Malt Type	Diastatic Power in Degrees Lintner
2-Row Pale Malt	110
6-Row Pale Malt	150
Pilsner Malt	125
Wheat Malt	120
Vienna Malt	100
Munich Malt	70
Pale Ale Malt	55

Diastatic Power of Common Malts

Specialty Malts

Specialty malts are treated differently from base malts, and this results in the development of different flavors and colors that enhance beer. Specialty malts are used to modify thickness, color, flavor, mouth-feel, and other characteristics. Some specialty malts add fermentable sugars, such as the lighter-colored crystal malts, but mainly they add unfermentable sugars for sweetness along with malty, caramel, and chocolate flavors.

Specialty malts can be broadly divided into two categories: crystal malts and roasted malts. Crystal malts are processed in such a way that the starches and proteins are mashed right inside the grain's hull. They are kept wet as the temperature is raised through the mashing range, and when they are dried, the sugars crystallize. A pale crystal malt has a lot of fermentable sugar and can be used almost like a malt extract. When crystal malts are heated to impart color, the darker they become the more of the sugars become unfermentable. So crystal malts will provide color, mouth-feel, and sweetness to beer.

Roasted malts are roasted or kilned to various degrees, resulting in differences in color and flavor. They contribute little or nothing in terms of fermentable sugar and with the exception of light Munich malts with a diastatic power of 70° Lintner, they have no enzymes. Each malt has its own profile due to differences in the variety of grain used to make it, and the techniques used in germination and drying prior to kilning or roasting.

Adjunct Grain

Adjunct grains are added to provide sugar, body, flavor, and enhanced head retention in beer. With the exception of instant oatmeal, they should be subjected to beta glucans and protein rests in mashing. In other words, anytime you use adjunct grains, you should utilize step-mashing procedures. Because these grains aren't malted, they should be used in conjunction with base malts that will contribute enough enzyme activity to convert the starches they contain to sugar.

When using adjunct grains, you want the total diastatic power of your mashed grains to average fifty or greater. If it is less than that, you won't get full conversion. You can assess the average diastatic power by use of the following formula:

((Pounds * Diastatic Power) + (Pounds * Diastatic Power) …)/ Total Pounds Grain

So if I have a recipe that includes two pounds of two-row malt, five pounds of pale malt, one pound of unmalted wheat and one pound of oatmeal, the average diastatic power of my grain bill is:

$$((2 * 110) + (5 * 55) + (1 * 0) + (1 * 0))/9 = 55$$

You can use any grain or even non-grains such as amaranth in beer, but the grain needs to be mechanically processed in such a way as to make it easily mashed. Usually, this means flaking. Flaked grains are soaked, flattened, and then dried. Because of the soaking process, their cell walls are already broken down. When using adjunct grains that have not been flaked, you need to cook them first by boiling them in water until tender.

In addition, when adjunct grains are included in the lauter, they lack the hulls needed for filtration so it is helpful to add rice hulls (the outer hulls of rice) to the mash in a volume roughly equal to that of the adjunct grain employed. This will keep the grains in the lauter tun from compacting so they can't filter. You can buy rice hulls at stores that cater to home brewers. A large bag sells for less than $2. Here is some information on some commonly-used adjunct grains:

Oatmeal comes in various forms. Instant oatmeal (the kind in the canister, not the packets) can be added directly to your grains. The old-fashioned kind needs to be prepared first (by following package directions) before being added. Oatmeal is a popular adjunct for higher alcohol dark beers, and it adds a smooth, creamy mouth-feel to beer. Use at 5% to 10% of total grains.

Flaked Rice is available at brewing stores. It is rice that has been crushed while wet and then dried. The flaking process makes it more easily mashed. Rice adds little if any flavor but its starches are readily converted, so it is good for raising the alcohol content of beer without adding a thicker mouth-feel. Use at 5% to 20% of total grains.

Flaked Barley is used in high alcohol beers to give body and a nice head. You can use from 5% to 10% flaked barley in your mash.

Wheat can be added to beer as either a malt (treat malted wheat as a base malt with a diastatic power of 120° Lintner) or as an unmalted adjunct grain. As an unmalted adjunct grain, it is flaked like rice or barley. It adds a lot of protein to beer and a pronounced starch haze along with distinctive flavor. Use anywhere from 5% to 20% flaked wheat in the mash tun.

Flaked corn is a popular adjunct grain with commercial breweries in the U.S. because it is plentiful and cheap. It adds a bit of flavor while lightening the color and increasing the alcohol content. Unfortunately, most of the corn available is genetically modified and there is no labeling requirement. You can use anywhere from 5% to 20% though once you use more than 10% you can expect a bit of corn flavor.

Adjunct Sugar

The amount of fermentable sugar made available to the yeast in your wort is a function of both the ingredients and the mashing temperature. Higher mashing temperatures yield fewer fermentable sugars and a thicker mouth-feel while lower mashing temperatures create a thinner beer with more fermentable sugars. The amount of fermentable sugar relates directly to the alcohol content of the finished beer, and alcohol content directly affects how well the beer is preserved. Therefore, it is common to add adjunct sugar to beer worts if the brewing ingredients and temperatures are likely to yield a beer with too little alcohol. When adjunct sugars are used, they are best added to the boil so they will be sterilized. Take the original gravity measurement after the boil so the sugar can be measured.

Beer yeast cannot directly utilize cane/table sugar. The cane sugar must be hydrolyzed into fermentable glucose and fructose first, and when yeast perform the hydrolysis enzymatically the process can add a cider-like taste and smell to beer. To get around this problem, make your own invert sugar. Invert sugar is table sugar that has been converted into glucose and fructose by a catalyzed reaction. To make invert sugar, put ¾ cup water, ⅛ teaspoon citric acid and ½ pound cane sugar in a pot on the stove. Bring to a light simmer and hold there for twenty minutes while stirring. Invert sugar is used widely in British and Belgian beer styles.

There are many other forms of fermentable sugar you can add to the wort, and the type of sugar used can affect the character of the beer. You can use honey, maple syrup, unsulfured molasses (the sulfured kind will hurt the yeast), brown sugar, demerara sugar, corn sugar, turbinado sugar, and more. Strongly flavored sources of sugar should be included at no more than 5% of the weight of the grain used, and more neutral sugars shouldn't exceed 10% of the weight of the grain used. So if you are using ten pounds of grain, you shouldn't use more than half a pound of molasses or a pound of honey in your batch.

Potential Extract

When creating your own recipes, one of the most important things you'll need to know about your grains is potential extract. Potential extract is a measure of how much of the weight of the grain or other adjunct will become dissolved in the wort. This affects the original gravity (OG). Potential extract is an absolute theoretical maximum, and in a home setup your extraction will be less efficient due to imperfect mashing, sugars not dissolved while lautering, and so forth. You can expect your efficiency to be somewhere between 80% and 90%, so for the design of recipes, we will use 85% efficiency.

The potential extract of grain, malt, or sugar is given in "points per pound per gallon" abbreviated as ppg. A "point" is an increase of .001 in specific gravity. You can predict the original gravity of your wort via the following equation:

Original Gravity = 1 + 0.85 * (Pounds Per Gallon * Potential Extract * .001)

You can also use the equation in reverse. If you know what you want your OG to be, you can figure out the pounds per gallon that you will need through re-arranging the equation as follows:

Pounds Per Gallon = (OG—1) / (.00085 * Potential Extract)

For example, if I want a beer with an OG of 1.070 and I am using a malt with a potential extract value of 32, the pounds of malt that I need per gallon of wort after boiling is:

Pounds Per Gallon = (1.070—1) / (.00085 * 32) or 2.6 pounds per gallon. So if I were making five gallons of beer, I would need 5 * 2.6 or 13 pounds of malt.

The following table lists the potential extract for a variety of common malts, adjunct grains, and sugars.

Malt or Adjunct	Potential Extract
American 2-row malt	37
American 6-row malt	35
Pale Ale Malt	34
Pilsner Malt	36
Munich Malt	33
Wheat Malt	39
Crystal Malts	34
Roasted/Kilned Malts	29–30
Dry Malt Extracts	45
Honey	35
Corn Sugar	40
Cane Sugar	45
Flaked Barley	32
Flaked Corn	37
Flaked Oats	37
Flaked Rye	36
Flaked Wheat	35

Potential Extract Value of Malts and Adjuncts

Color of Beer

Like malt, the color of beers is expressed in degrees Lovibond, a measure that also corresponds to the more scientific version called Standard Reference Model or SRM. SRM is measured using a spectrophotometer, and thankfully such devices aren't really needed. For purposes of predicting the color of beer from its ingredients, just two equations are necessary.

First, the Malt Color Units (MCUs) are calculated for each grain, and the MCUs are added together.

MCU = (Pounds of Grain * Grain Color in Degrees Lovibond) / Volume of Beer Batch in Gallons

Then, the added MCUs are converted to SRM with the Morey equation, for which you'll need a scientific calculator:

SRM = 1.4922 * MCU0.6859

For example, if I am making five gallons of beer from eight pounds of pale ale malt at an average of 2.2 degrees Lovibond, and one pound of crystal malt at an average of 62 degrees Lovibond, the math would look like this:

MCU (Pale Ale Malt) = (8 * 2.2) / 5 = 3.52

MCU (Crystal Malt) = (1 * 62) / 5 = 12.4

MCU (Total) = 3.52 + 12.4 = 15.92

$SRM = 1.4922 * 15.92^{0.6859} = 9.96$

For some points of reference, Miller Lite is 2 SRM, Bass Pale Ale is 10 SRM, Porter is 29 SRM, and Imperial Stout is 70 SRM.

Hops

Beer, being made from malt, can be a relatively sweet beverage. Without something to balance the sweetness, it isn't a very appetizing drink. The sweetness of hot chocolate is balanced with the bitterness of cocoa, the sweetness of cola drinks is balanced with the sourness of phosphoric or citric acid, and the sweetness of beer is balanced with the bitterness of hops.

The more malt that a beer contains—or, more properly, the more unfermentable sugars imparted to a beer by its malt—the more bitterness it needs as a balanced beverage. A beer can be made completely from malt and require very little balancing bitterness because the malt was so efficiently turned into fermentable sugars that little residual sweetness remains. Most of the pale lagers mass-marketed in the United States have little residual sweetness, so they don't require much hopping, whereas the dark heavy brews of Britain require considerable hopping.

Hops are used either for bittering or for flavor/aroma. Different hop varieties have been developed that are better for each purpose, though to some degree there is interchangeability. Hops used for bittering have a high percentage of alpha acids, whereas those used for flavor/aroma have a much lower percentage. The thing is, hops are expensive, so even though you could use flavor hops for bittering, it is more economical to use a bittering hop for that purpose because the higher alpha acid content allows less of them to be used.

In addition, breeders pay no attention to the flavor and aroma

◆ Bittering and aroma hop varieties.

characteristics of bittering hops because when they are boiled in the wort, all of the flavor and aroma are driven off. As a result, the flavor and aroma of bittering hops is usually inferior to that of hops bred for the specific purpose of providing aroma and flavor.

Hops are extremely vulnerable to spoilage. As a result, most of the hops available are sold as compressed plugs in nitrogen-filled sealed packages that are refrigerated. The following table lists some readily available varieties of hops along with pertinent information. Dual-use hops are noted by listing both uses, in the order of suitability.

Variety	Purpose	Alpha Acid %	Description
Bullion	Bittering	5–10	Bitter, pungent
Cascade	Aroma	4–7	Floral, spicy
Centennial	Aroma/Bittering	9–11	Floral, spicy
Challenger	Aroma/Bittering	7–10	Complex floral
Chinook	Bittering	11–14	Bitter, fragrant
Cluster	Bittering	5–8	Non-distinctive, bitter
Comet	Bittering	9–11	Bitter
Crystal	Aroma	2–5	Spicy
Fuggles	Aroma	4–6	Spicy, fragrant, mild
Goldings	Aroma	4–6	Spicy, fragrant, mild
Haulertau hersbrucker	Aroma	4–6	Fresh, floral
Liberty	Aroma	4–5	Spicy, mild
Magnum	Bittering	12–14	Bitter, clean
Northern Brewer	Bittering	7–11	Complex, bitter
Nugget	Bittering	11–16	Bitter
Progress	Aroma	6–7	Fragrant
Spalter	Aroma	4–5	Spicy
Williamette	Aroma	4–6	Spicy, mild
Yeoman	Bittering	9–14	Heavy

Hop Usage Chart

Determining the Amount of Hops

The weight and type of hops needed for a particular beer are dictated by the style of the beer or the recipe if you are following such guidelines. But if you are designing your own recipe, you can calculate the bittering hops needed to balance the maltiness of the beer using some equations and a scientific calculator.

The bitterness of beer is measured in IBU, International Bitterness Units. Each established style of beer has a range of IBU for that particular style. However, we are not duplicating styles per se, because there is already plenty of data on that topic. Instead, we are trying to determine the range of hopping needed for any beer we happen to make.

In general, the higher the original gravity of the beer, the more bitterness it needs in order to counter residual sweetness from unfermentable sugars. When making your own beers, you may find the guidelines in the following table helpful.

Original Gravity	Bitterness in IBU
1.030	6–12
1.040	10–16
1.050	14–24
1.060	20–32
1.070	30–40
1.080	36–48

Balanced Bitterness Range by Original Gravity

The hops added for flavor/aroma during the final five minutes of the boil or added directly to the secondary fermenter (a practice known as dry hopping) contribute little if any bitterness. They are selected on the basis of their flavor and aroma profile, and you can use one ounce for a five-gallon batch in all cases. From there you can experiment with using a little more or less, and different varieties.

Most of the IBUs in beer come from the hops that are added for bitterness. Bittering hops are added at the start of the sixty-minute boil. The question is: How many ounces of hops do you add to achieve a certain level of bitterness? This is determined by the alpha acid content of the variety of hops you have chosen and the gravity of the wort, because the higher the gravity of the wort, the more compounds have already been dissolved, so the less efficiently the isomerized alpha acids will be dissolved. Glenn Tinseth[20] did a tremendous amount of work that culminated in some very useful equations that yield data that is close enough for sound recipe design, and their result after algebraic manipulation for a five-gallon batch is as follows:

Ounces of Hops = (6.7 * Desired IBUs) / (Alpha Acid percent * Utilization percent)

Because we are assuming a sixty-minute boil (boils longer than that make so little difference in utilization as to be undetectable), the factor that will determine utilization percentage is the specific gravity of the wort. Keep in mind that the specific gravity of the wort when first added to the brew kettle is going to be lower than at the end of the boil, because about half a gallon of water in the wort will be lost in steam. So if you have planned a recipe with a specific original gravity in mind, subtract five points when using the following table:

Wort Gravity	Percent Utilization
1.030	27.6%
1.035	26.4%
1.040	25.2%
1.045	24.2%
1.050	23.1%
1.055	22.1%
1.060	21.1%
1.065	20.2%
1.070	19.3%
1.075	18.4%
1.080	17.6%

Hop Utilization Factor for a 60-minute Boil

For example, if I am making a batch of beer with an original gravity of 1.065, how many ounces of magnum hops should I include in the boil? Looking at the chart relating bitterness to gravity, a bitterness of around 32 IBU should be fine. The hop chart shows that magnum hops are 13% IBU on average. The hop utilization chart shows my alpha acid extraction will be 20.2% efficient with a gravity of 1.065, but that will actually start a bit more diluted, so I will go with the 21.1%

20 Tinseth, Glenn (1995—1999) The Hop Page at http://www.realbeer.com/hops

efficiency correlated with a wort gravity of 1.060. Substituting these numbers into the equation, I get:

$$\text{Ounces of Hops} = (6.7 * 32) / (13 * 21.1) = .78 \text{ ounces.}$$

I can measure that out on my gram scale by multiplying the number of ounces (.78) by the number of grams in an ounce (28.3) to get 22 grams. So I add 22 grams of magnum hops at the beginning of the sixty-minute boil.

Yeast

Even with all other ingredients being identical and subjected to identical techniques, the strain of yeast used to make beer will have an effect on the beer's taste and character. Every strain of yeast works best within certain temperature ranges, and they will all have a different ability to turn sugars into alcohol. Different yeasts also create a different flavor profile in terms of esters, higher alcohols, and other organic products.

Because the special temperature requirements of lagering aren't available in most homes, I am going to concentrate on ale yeast. If you decide to start making lagers and have set up an appropriate area, the same principles apply. The strains I recommended earlier, Safbrew S-33, Wyeast 1056 and Wyeast 1084 are good all-purpose yeasts for making ale, but others will give you different results worth exploring. This list is by no means exhaustive, but rather represents strains that have worked well for me and my friends.

One aspect of the following table that may be useful is selecting a yeast that is well-adapted to the temperatures you can most easily maintain in your fermenting area. For most of the winter, for example, the temperatures I can maintain are between 60 and 65 degrees, which rules out some yeasts altogether but makes others a good choice.

Beer yeast is not as well adapted to high gravity wort as wine yeast. Hence, the higher the specific gravity of the wort, the more yeast you will need. Unless you are using a brewer's yeast that is specifically intended for high-gravity worts such as White Labs' WLP099, you will need to either use more packets of yeast, or amplify your yeast before pitching. As a general rule, you should add the same amount of yeast you would normally use for every ten points of gravity above 1.050. So if your gravity is 1.070, you will need three packets, but if it is 1.050 you will only need one.

I have successfully fermented worts with a gravity exceeding 1.050 without resorting to additional yeast, but it's not something I would count on. Given the

Name	Manufacturer	Attenuation	Temperature	Description
Abbey Ale, WLP530	White Labs	75–80%	66–72	Fruity
American Ale, 1056	Wyeast	73–77%	60–72	Balanced, dry, soft
American Ale II, 1272	Wyeast	72–76%	60–72	Clean, nutty, tart
BedfordBritish, WLP006	White Labs	72–80%	65–70	Traditional English style
Belgian Ale, 1214	Wyeast	72–76%	58–68	Estery, crisp
Belgian Ale, WLP550	White Labs	78–85%	68–78	Spicy
British Ale, 1098	Wyeast	73–75%	64–72	Dry, crisp, and fruity
British Ale, WLP005	White Labs	75–80%	68–75	Malty
Coopers Homebrew	Coopers	75–80%	68–80	Clean
Dry English Ale, WLP007	White Labs	70%–80%	65–70	Ferments high gravity ales to dryness
Fermentis US 56	Fermentis	77%	59–75	Clean and mild
German Ale, 1007	Wyeast	73–77%	55–68	Dry and mild
Irish Ale, 1084	Wyeast	73–77%	60–72	Fruity, complex
Irish Ale, WLP004	White Labs	73–80%	65–70	Fruity, dry, crisp
Muntons Standard	Muntons	80%	57–77	Clean, balanced
Nottingham	Danstar	75%–80%	57–70	Fruity
Safbrew S-33	Fermentis	75%	59–75	Good for high-gravity beers

Applications of Popular Yeast Varieties

time and effort involved in preparing beer, it's far better to be safe than sorry with something so simple as amplifying yeast. Yeast, especially liquid yeast, is pretty expensive, so it is better to amplify it yourself than buy additional packets. You can do the same with dry yeast, but the cost savings are not as great.

A further problem is that yeast doesn't live forever in the vials or packets in which it comes. Its viability—that is, the quantity of active yeast cells—declines with age. Dry yeast is less susceptible to this than liquid yeast, but any form of yeast that is within three months of its expiration date should be multiplied before pitching.

Multiplying yeast is easy. You make a low-gravity sterile wort, aerate it, pitch the yeast into it, keep it at room temperature, and let it multiply (shaking gently every few hours) for 18-24 hours before pitching. The primary variables affecting how much the yeast will multiply are the amount of wort used for propagation, the size of the starting colony, and the presence of adequate nutrition.

For yeast that is simply old, you can use sixteen ounces of sterile wort. For multiplying by a factor of two, use a quart. For a factor of four, use half a gallon.

To make sterile wort, clean out your kitchen sink and fill the basin with ice water. Put a pot on the stove containing as much water as needed for the amount of yeast you intend to propagate. Bring the water to a boil, and add 3½ ounces of dry malt extract (any brand will do) to the boiling water. Boil for fifteen minutes, then cover the saucepan and put it in the ice bath to cool until it reaches a temperature below 80 degrees.

Meanwhile, prepare the vessel that will hold your multiplying yeast by sanitizing it along with a stopper and airlock. I use a one-gallon jug. You can sanitize with either sulfite or Star-san for the job, just make sure you don't rinse after.

Once the temperature of the sterile wort is below 80 degrees but above 70 degrees, pour the sterile wort into your vessel, add the yeast, and install the sanitized stopper and airlock. Set aside for 18-24 hours, giving it a gentle swirl every few hours. When the time comes for pitching, give it a swirl and pour it smoothly into your primary fermenter.

Putting it Together: Beer Recipe Creation

The past couple of chapters have probably seemed endless, but here is where we bring it all together into something practical: a real-world recipe. I have on hand two pounds of Pilsner malt, five pounds of crystal malt, and ten pounds of pale ale malt. I want to use up the Pilsner malt and make a dark amber ale. I also have some cascade and fuggles hops on hand. I will use Wyeast 1084 liquid ale yeast because its temperature range fits well with my dining room in winter. I'd like a fairly high alcohol content so the beer keeps pretty well because I'd like to give it as a gift.

Two parts of our grain bill have already been decided: two pounds of Pilsner malt and two pounds of dark crystal malt for color. I'd like a starting gravity of about 1.065, so how much pale ale malt will I need for the five-gallon batch?

Here it would be helpful to go back and review the section on potential extract. The amount of Pilsner malt per gallon is two pounds in five gallons, so ⅖ = 0.4. The potential extract of Pilsner malt is 36 and my efficiency is 85%, so the Pilsner malt gives me 0.4 * 36 * .85 = 12.2 points of gravity.

The amount of crystal malt per gallon is two pounds in five gallons, so ⅖ = 0.4. The potential extract of crystal malt is 34 and my efficiency is 85%, so the crystal malt gives me 0.4 * 34 * .85 = 11.6 points of gravity.

So my existing ingredients have already given me 11.6 + 12.2 = 23.8 points of gravity. I need 65 points of gravity total, which means I need to use the pale ale malt to give me 65−23.8 = 41.2 points of gravity.

Pale ale malt has an potential extract of 34, and my efficiency is 85%, so the amount of pale ale malt that I need per gallon is 41.2 / (34 * .85) = 1.4 pounds. I am making five gallons, so I will need 5 * 1.4 pounds for a total of 7 pounds of pale ale malt.

So my grain bill for mashing is as follows:

7 pounds pale ale malt
2 pounds Pilsner malt
2 pounds crystal malt

I will be using Wyeast 1084 yeast, which has an average attenuation of 75%, so my final gravity will be: 1.065 - .75(1.065 -1) = 1.016. The alcohol in my beer will be given by the equation given in the previous chapter: Alcohol Percent by Volume = 117 * ((OG−1)−(FG−1)).

117 * ((1.065−1)−(1.016−1)) = 5.73%

I would like a well-hopped beer. I have some cascade hops with an average alpha acid percentage of 7%. My original gravity is planned to be 1.065, but it will be a bit more diluted when I first put it in the brew kettle, so I'll use the utilization percentage for 1.060, which is 21.1%.

The equation for the amount of hops to use is:

Ounces of Hops = (6.7 * Desired IBUs) / (Alpha Acid percent * Utilization percent)

so

(6.7 * 32)/(7 * 21.1) = 1.45 ounces of hops.

So my recipe requires 1½ ounces of cascade hops, and I'll use one ounce of fuggles hops in the last five minutes of the boil for aroma.

My recipe so far looks like this:

7 pounds pale ale malt
2 pounds Pilsner malt
2 pound crystal malt
1.5 ounce cascade hops (bittering, at the beginning of boil)
1 ounce fuggles hops (aroma, at the end of boil)
1 packet of Wyeast 1084 Irish Ale yeast

What else does my recipe need? All that is left to add are any clarifying agents and water treatments. My bottle of Irish moss has directions right on the package: one teaspoon per five-gallon batch during the last fifteen minutes of the boil. That was easy. But what about the water? My own water is so acidic that it leaves green residue from the copper it dissolves in the pipes. It's not drinkable, so it would make horrible beer. Instead, I will use bottled water from Monadnock Mountain Spring Water and support my local business.

Referring back to the chapter on the science of beer, the water analysis on their website indicates the water contains practically nothing, but my mash water requires at least 50 ppm (also known as mg/L) of calcium and 10 ppm of magnesium. Using the table in the section on water chemistry, I decide to add three grams of magnesium sulfate for 15 ppm of magnesium and 60 ppm of sulfate, three grams of sodium bicarbonate for 45 ppm of sodium and 114 ppm of carbonates, and four grams of calcium chloride for 56 ppm of calcium and 100 ppm of chloride.

So my final recipe looks like this:

Monadnock Dark Amber Ale

7 pounds pale ale malt (2.2 Lovibond average)
2 pounds Pilsner malt (1.7 Lovibond average)
2 pounds crystal malt (62 Lovibond average)
1.5 ounce cascade hops (bittering, at the beginning of boil)
1 ounce fuggles hops (aroma, at the end of boil)
1 packet of Wyeast 1084 Irish Ale yeast
1 tsp Irish moss
3 g Magnesium sulfate (Epsom salts)
3 g Sodium bicarbonate (baking soda)
4 g Calcium chloride
We can now calculate its color as well.
MCU (Pale Ale Malt) = (7 * 2.2) / 5 = 3.08
MCU (Pilsner Malt) = (2 * 1.7) / 5 = .68
MCU (Crystal Malt) = (2 * 62) / 5 = 24.8
MCU (Total) = 3.08 + .68 + 24.8 = 28.56
SRM = 1.4922 * 28.560.6859 = 14.86

This beer will be a nice dark amber. Because all of the malts in this beer are well-converted, I will use a single-infusion mash. Because I'd like it to keep well, I will increase the alcohol content by mashing at temperatures between 149 and 153. Since it is an ale, it will require from three to six days of primary fermentation, and I'll condition it in secondary fermentation for six to twelve days before bottling. Lastly, because it is a high-gravity wort, (its specific gravity is higher than 1.050), we'll also need to amplify our yeast.

Now that we have our recipe, let's proceed to the next chapter where all of the foregoing information will be practically applied!

10

Brewing Techniques

My homemade Monadnock Dark Amber Ale is made from inexpensive ingredients I happen to have on hand. It is far from the only recipe you could make. Using the tables and principles in the foregoing chapters, you could make an infinite variety of beers. But for purposes of illustration, this is a very typical beer, and given how little it costs in terms of ingredients, you'll be impressed with how excellent it is.

From Grain to Primary Fermenter

1. **Clean your mash** tun thoroughly, then put your grains in the mash tun and mix them up.

2. We are using 11 pounds of grain. You use between one and one and a half quarts of water per pound of grain when mashing. There are four quarts per gallon, so I am heating three gallons of water. The water poured into the mash tun is called the strike water. I want the water in the mash tun to be between 152° and 158°. Closer to 158 gives more mouth-feel and closer to 152 gives more alcohol. I will split the difference and aim for 155. Because the grain is at room temperature, this will cause the temperature of the water I add to drop. Therefore, I will be heating the strike water to 165. While it is heating, I will add the 3g magnesium sulfate, 3g sodium bicarbonate, and 4g calcium chloride.

3. Pour in your strike water and immediately stir thoroughly, and check the temperature. If it is too high or low, correct with the addition of colder or hotter water.

4. Check the pH. If it is above 5.8, add phosphoric acid. If it is below 5.2, add potassium hydroxide solution. Additions should be 1 cc at a time if you have to bring the pH up or down more than a point, and ½ cc at a time for a half a point.

5. Allow to sit for an hour covered to preserve heat. At the end of an hour, use an iodine test for starch conversion. If the test

⊗ Mixed grains in the mash tun.

⊗ Pouring the strike water into the mash tun.

⊗ After adding a bit of potassium hydroxide, the pH is in the proper range.

⊗ Using iodine to test for starch conversion.

shows inadequate starch conversion, allow to sit another half hour then test again.

6. Get your sparge water ready. Heat three gallons of water on the stove to 170 degrees. Test the pH. Adjust the pH to 5.4 to 5.6 using either phosphoric acid or potassium hydroxide solution as appropriate.

⊗ Pouring the converted mash into the lauter tun.

7. When the mash has converted and the sparge water is at 170 degrees, pour the contents of the mash tun into the lauter tun.

8. Run a couple of quarts through the lauter tun into a pitcher and pour back into the top until the lauter tun is filtering adequately. It should only take a couple of times.

9. Break the yeast packet and put in a warm place. I keep it under my shirt.

⊗ Use a pitcher to pour the first couple of quarts back through the lauter tun.

10. Run your lauter slowly into your brew kettle, adding sparge water to the lauter tun a little at a time as the wort drains into the brew kettle. Check the specific gravity of the wort coming through the lauter tun from time to time. Stop when the specific gravity reaches 1.008.

11. Bring your brew kettle to a boil, set a timer for sixty minutes and add your bittering hops.

⊗ Running the wort from the lautering tun into the brew pot.

12. While the wort is boiling, clean and sanitize (I use Star San, a no-rinse sanitizer) your primary fermenter, and put the hose adapter on the faucet for your wort chiller.

13. With 15 minutes left in the one-hour boil, add the Irish moss to the boiling wort, and insert your cleaned wort chiller. (I also douse my wort chiller with Star San in the primary fermenter as I'm cleaning the primary fermenter.)

⊗ Immersing the wort chiller.

Brewing Techniques

⊗ Using the wort chiller to cool the wort.

⊗ The specific gravity is exactly as we planned!

⊗ Pitching the yeast.

14. With 5 minutes left in the boil, add your aroma/flavor hops.

15. The second the boil is over, move your brew kettle with the wort chiller onto a chair or similar holder, attach your wort chiller to your sink and turn on the cold water.

16. Monitor the temperature of your wort. Run the wort chiller until the temperature is between 70 and 80 degrees. (Monitor the connections to your wort chiller for leaks and tighten the hose clamps if needed.)

⊗ The primary fermenter is filled and ready.

17. Remove the chiller, pour the wort into the primary fermenter and aerate the wort by splashing it around. Alternately, you can use a sanitized racking cane with an aerator attached to the end of the hose that goes in the fermenter.

18. Use your wine thief to get a sample of wort so you can check and record your original gravity. In this case, the OG is exactly as planned.

19. Use sulfite solution to sanitize the expanded yeast packet, then open it carefully and pitch the yeast by pouring it gently into the wort.

20. Affix the sanitized lid and insert an airlock filled to the fill line with sulfite solution. Carefully, without splashing, put the fermenter wherever it should be. Because my floors are cold in winter when I make beer, I put my fermenter on a table.

Now . . . you wait. Within twelve hours and definitely in less than a day, you should start seeing bubbles issuing from the airlock. If you sniff the airlock, it will smell like beer. The primary fermentation will continue for three to six days depending on the temperature, the nutrients in the wort, and the strain of yeast. You'll know primary fermentation is complete when the rate at which bubbles issue through the airlock has reduced to one or two per minute. If you are busy, you can leave the beer in the primary for as long as another week, but you don't want to leave it too long or the yeast and debris that has accumulated at the bottom of the fermenter will start to break down and impart bad flavors to the beer.

Once primary fermentation is finished, I prefer to condition ales through a period of secondary fermentation that is twice as long as the primary fermentation, with a minimum of one week and a maximum of three weeks. I also take that opportunity to add finings such as gelatin and polyclar to help clear the beer. In this case, we'll use a one-week secondary fermentation, with finings added at the time of racking.

Racking Your Beer to the Secondary Fermenter

If you read the chapters on wine, this process is identical and uses identical gear, except we'll be adding a clarifier to the beer and sterilizing any top-up water because the alcohol level in beer is low enough to make it more vulnerable to spoilage.

1. Put ½ gallon of water in a pot, boil it for 15 minutes then turn off the heat and cover the pot. Let it cool to 80 degrees or less.
2. Use your scale to measure out one gram of fining gelatin, and mix that with two tablespoons of cold sterilized water in a sanitized coffee cup. Separately, put seven tablespoons of water in a glass measuring cup, and heat on high in the microwave for one minute. Add the hot water to the dissolved gelatin in the coffee cup, mixing thoroughly. Allow this to cool down to a temperature of 80 degrees, add ¼ teaspoon of Polyclar to the gelatin mixture. Cover with a sanitized cover until ready for use.
3. Thoroughly clean a glass carboy and then sanitize with either sulfite solution or a no-rinse sanitizer such as Star San. Do the same thing

❯❯ Sanitizing the secondary fermenter.

≫ Racking from the primary fermenter into the secondary fermenter.

with your auto-siphon/racking cane and the plastic hose, the rubber stopper for the carboy, and an airlock.

4. Put the hose in the secondary fermenter, gently lower the auto-siphon into the wort in the primary fermenter, and give it a pump or two to get the flow started.

5. Continue until either the carboy is full with only minimal airspace or the auto-siphon is sucking up gunk. Then stop. Pour some of the prepared fining gelatin into the carboy at intervals until it is all used up.

6. If there is still room in the primary fermenter, top it off with water you boiled earlier.

7. Apply the rubber stopper and an airlock filled with water to the fill line.

8. Use a carboy handle and gently move the carboy somewhere at a moderate 60-70 degree temperature that is out of the way. If sunlight enters the area, cover the fermenter with a blanket to exclude light.

≫ This carboy needs a little boiled water added to reduce the airspace.

There is no precise point at which secondary fermentation can be said to be completed, because it is primarily a bulk aging rather than fermentation process. Because beer has a relatively low alcohol content compared to wine and a lot of dissolved solids in comparison as well, it isn't suitable for long-term aging.

For ales the period of secondary fermentation can range anywhere from one to three weeks. A good rule of thumb is it should be bulk aged for somewhere between as long as the primary fermentation took and twice as long, with a three week maximum. For lagers, because the aging takes place at just above freezing so biological processes are slower, you can bulk age for as long as six weeks.

In this case, we will bulk age for a week and then bottle. Bottling is likely the least fun aspect of home brewing simply because it is repetitive. For every five gallons of beer, you are filling as many as 53 bottles, so the process can be a bit tedious. On the bright side, once the beer is bottled, it will be sufficiently carbonated and bottle-aged to be consumed in two weeks! Nothing is better than beverages you've made yourself.

Bottling your Beer

The single most important aspect of bottling beer is sanitation. The second most important aspect of bottling beer is also sanitation. Did I mention sanitation? In all seriousness, this is where your beer is most vulnerable to contamination, so cleanliness is important.

1. Sanitize your bottling bucket and your primary fermenter bucket. You can use sulfite solution mixed with three tablespoons of sulfite powder and one teaspoon of citric acid per gallon of water, or a no-rinse product such as Star San mixed according to label directions. Make at least 2 gallons of sanitizing solution in each bucket.

⊗ Sanitizing the beer bottles.

2. Immerse your racking tube, bottling wand, and the lengths of plastic hose you'll be using in the sanitizer in your primary fermenter bucket and pump sanitizer throughout the entire works. Leave everything to soak.

3. Clean your beer bottles thoroughly, and then immerse them in sanitizing solution in your bottling bucket. You won't be able to fit all the bottles in the bucket. Leave them filled with sanitizing solution for five minutes, then dump the solution out of the bottles back into the bottling bucket and set the bottles aside. Repeat until all beer bottles have been cleaned and sanitized.

⊗ Sanitizing the bottle caps.

4. Put two pots on the stove. In one pot, put 50–60 bottle caps in water, bring the water to a boil for several minutes, then turn off the heat and put the lid on the pot to prevent contamination. In the other pot, bring two cups of water and five ounces of corn sugar to a boil, then turn off the heat and cover. This is your priming sugar and you can use a greater or lesser amount depending on the desired degree of carbonation. (See the chapter on the science of beer for details on how much sugar to use.) Five ounces of corn sugar gives a level of carbonation equivalent to most commercial beer.

⊗ The priming sugar solution is cool enough to add to the bottling bucket.

5. Place your secondary fermenter on a table or counter. You may wish to put a short wedge under the edge of the fermenter furthest from the edge to help get all the beer out. Empty all the sanitizing solution out of your bottling bucket, but do not rinse it. Place your bottling bucket on the floor in front of your secondary fermenter.

6. Place your pot of sugar solution in an ice bath in the sink to cool until the temperature of the solution is between 70 and 80 degrees. Then pour the solution into the bottom of the bottling bucket.

7. Take your racking tube and the plastic hose for it out of the sanitizing solution in the primary fermenter bucket, and empty out as much sanitizer as you can.

8. Carefully, so as to avoid disturbing sediment, use your racking tube to transfer the beer from the secondary fermenter to the bottling bucket. Do not let the beer splash around because it is vulnerable to oxidation at this stage.

⊗ Transferring the beer from the secondary fermenter to the bottling bucket.

9. Set aside the empty secondary fermenter to be cleaned, and gently lift your bottling bucket full of beer onto the counter or table. Put your bottles on the floor in front of the bottling bucket. To avoid spills, I put my sanitized bottles open-end up in the cases.

⊗ Filling bottles using the bottling wand is easy.

10. Remove the bottling wand and the plastic hose from the sanitizing solution in the primary fermenter, and empty as much sanitizing solution from them as you can. Use sanitizing solution on a paper towel to wipe down the nipple on the spigot of the bottling bucket, then affix the hose to the nipple and the bottling wand. Open the valve.

11. The bottling wand is made with a spring-loaded valve that allows liquid to flow when you press down on it. Put the wand into a bottle, press against the bottom of the bottle, and ease up as the beer reaches the top of the bottle's opening. When you withdraw the bottling wand, a bit of airspace will be left at the top. This space is called ullage.

12. Repeat until all bottles are filled. Then, without touching the part of the caps that will contact the beer, lightly place the caps on all the bottles. Leave for about ten minutes so that gasses from the beer can displace the oxygen.

⊗ Using the capper to crimp the caps in place on the beer bottles.

13. Use your bottle capper to crimp the caps into place. Use slow, steady pressure. If you mess up, don't be concerned. Just discard the cap and use another. That's why you boiled extras!

14. Once your beer is bottled, clean and sanitize everything. Then let it dry, and put it away. Once the equipment is dry you might even consider keeping it in the large black plastic trash bags to keep dust and debris out.

15. Keep your beer at room temperature for two or three weeks.

16. Enjoy and share with your friends!

⊗ A nice label enhances the perception of the quality of your product.

17. Remember, it is legal to share and even give away homebrew to your friends and neighbors, but it is highly illegal to sell it or distill it without a license.

Labeling

Labeling your finished beer serves two purposes: identification and presentation. Once you have been making beer for a while, you will have bottles floating around of various origins, and labeling the beer will be important just to know what you have.

Presentation is everything. People are prejudiced regarding the quality of what is in the bottle by the fashion in which it is presented. A nice label makes all the difference in creating a positive prejudice. Though there are places on the Internet that will make custom labels for you, in some cases they cost over a dollar each. Obviously, that's more expensive than the beer itself and should be reserved for beer being presented as a gift or for a special occasion. The rest of the time, perfectly presentable labels can be created on label stock using an inkjet printer.

Logs

This is probably starting to get old, but I'll continue to say it about everything: keep a log of your beer-making efforts. Your log is an invaluable tool for improving your product and learning to get the exact results you need. It is only because of my logs that I was able to ultimately derive the simple formulas and recipes in this book.

With beer, very subtle differences in technique—such as a mashing temperature of 152 as opposed to 155—can make a big difference in the results. Make note of everything and by the time you are on your fifth batch of beer, you will start to appreciate the value of your log. It will also help you duplicate particularly good batches of beer you've made in the past.

PART IV

Vinegar from Beer and Wine

Markham Farm Private Reserve
283 Turnpike Road
New Ipswich, NH 03071

Apple Wine

- 2007 -

50/50
Cyser
Mix

11

Principles and Materials for Vinegar

There is vinegar, and then there is vinegar. Most often, we buy vinegar as a commodity product without giving much thought as to quality. The gallon jugs of distilled vinegar in the supermarket are indistinguishable. There is no point in making your own vinegar when you can buy it for $1/gallon in bulk; so this chapter is not about making that kind of commodity product.

Really good vinegar is a complex taste sensation to be savored and appreciated. It takes on the character of the malt, cider, or wine from which it is derived. It can also be improved by aging as the complex flavor and aroma compounds meld, recombine, and change. It is truly a gourmet product, and hand-crafted examples are usually more than $20/pint.

Vinegar in general is a healthy condiment. Vinegar increases satiety[21] thereby reducing caloric intake, it reduces the glycemic index of foods with which it is consumed,[22] and may reduce the risks of certain types of heart disease.[23] And just as wine preserves many of the vitamins and antioxidants in the original fruit, homemade vinegars made from those wines will likewise preserve vitamins and antioxidants; thereby making it even more healthy than the commodity vinegars used in the studies.

So this chapter is not about duplicating commodity products that are cheaper to buy than they are to make. Rather, it is about making a uniquely healthful product with gourmet qualities that will enhance your salads, greens, dressings, and anything else you make with vinegar.

If you make wine and beer, you will already have the raw materials at hand allowing you to make gourmet vinegar inexpensively. So I will focus on using wine and beer as the starting materials in this chapter, even though vinegar can also be made using similar techniques if you use hard cider, sake, or practically any other product containing alcohol.

Speaking of Wine and Beer

I really enjoy making wine and beer. I enjoy every aspect of the process, and I especially enjoy sharing my work with someone who will appreciate the results of my efforts. But sometimes my efforts result in a less-than-stellar product. The wine I made from bottled blueberry juice and brown sugar comes to mind as does the beer I made with far too much oatmeal. What on earth was I thinking? The good news is, I can use these to make vinegar. You will likely have some learning experiences of your own that will serve as excellent raw material.

Some commercial wines and beers are pretty poor. Even if wine or beer that you've purchased is pretty good, it may have sat in the refrigerator too long or be

---

21 Östman, E; Granfeldt, Y; Persson, L; Björck, I (2005). "Vinegar supplementation lowers glucose and insulin responses and increases satiety after a bread meal in healthy subjects." *European Journal of Clinical Nutrition* 59 (9): 983–8

22 Johnston, C. S.; Kim, C. M.; Buller, A. J. (2004). "Vinegar Improves Insulin Sensitivity to a High-Carbohydrate Meal in Subjects With Insulin Resistance or Type 2 Diabetes." *Diabetes Care* 27 (1): 281–2

23 Johnston, Carol S.; Gaas, Cindy A. (2006). "Vinegar: medicinal uses and antiglycemic effect." *MedGenMed* 8 (2): 61

near its sell-by date. Rather than dump that effort or money down the drain, you might consider using it to make your own vinegar.

Wine that you use to make vinegar cannot have been preserved using potassium sorbate or sodium benzoate. Beer seldom has such preservatives. Any wine you use can be normally sulfited or it can be non-sulfited. The wine can be white or red, sweet or dry, and made from any conceivable edible fruit. The beer you use can be made from barley, wheat, or any other grain. And even though you will likely choose wines or beers for this process that were not optimal for drinking, it is very important that the starting material you choose be biologically sound.

Any sound wine or beer that you use, even if it isn't very good for drinking, will still yield a product far superior to the "wine vinegar" or "malt vinegar" you will find at the supermarket. The "wine" they use as a starting product was never intended for drinking in the first place, whereas yours was planned with drinking quality in mind and is hence a better material from which to make vinegar.

What is Vinegar?

Vinegar is a dilute form of acetic acid, ranging in strength from 4% to 8%. It is made by the oxidation of ethyl alcohol into acetic acid through a fermentation process undertaken by acetic acid bacteria (AAB). Just as the yeast in wine derives its energy from sugar and produces ethyl alcohol as a waste product, AAB derive their energy from alcohol and produce acetic acid as a waste product. And just as the ethyl alcohol in wine acts as a preservative against organisms that cannot tolerate alcohol, acetic acid acts as a preservative against organisms that cannot tolerate the low pH created by acetic acid. This is how pickling foods in vinegar keeps them from spoiling.

Acetic Acid Bacteria

There are a great many specific strains of AAB. They are present on the surface of both healthy and damaged fruit as well as the nectar of flowers. They are also commonly transferred by the fruit flies that could have been attracted to your wine-making or brewing process.

Wine is produced in anaerobic conditions, meaning that oxygen is excluded. Vinegar, on the other hand, is produced under aerobic conditions as the AAB require oxygen to work. In the absence of oxygen, the bacteria go dormant.

Various strains of AAB[24] are present in wine must from the very beginning and remain in the wine even when it is bottled.[25] The primary factor that keeps it suppressed in wine is lack of oxygen and alcohol levels that are too high for the bacteria to process. So especially with newly-made wines, all that is theoretically necessary to turn wine into vinegar is to permit the entry of oxygen. In the presence of oxygen the bacteria would quickly proliferate as a film on the surface of the wine and turn the alcohol to acetic acid, especially if the alcohol level is under 10%.

Beer is even more susceptible to acetification because its lower alcohol content, lack of sulfites, and higher nutritional content make it an attractive target.

Acetic acid bacteria are not the only bacteria that can take hold in wine or beer, and leaving the results to chance can result in a product that is not only unusable, but thoroughly rotten. So for our purposes, just as a specific strain of yeast is used to make wine, a specific strain of bacteria is used to make vinegar. Acetic acid bacteria are commercially available in a form called vinegar mother. Vinegar mother, also known as *Mycoderma aceti*, is a gelatinous substance containing the AAB that forms on the surface of vinegar. Though vinegar could certainly be made from *Gluconobacter oxydans* or *Acetobacter pasteurianus* among many other possibilities, all of the commercially available vinegar mothers are *Acetobacter aceti*.

Acetobacter aceti needs to float on top of the wine or beer you use to make vinegar so that it has access to oxygen at all times. Without access to oxygen, it will go dormant. The vinegar mother you obtain may look like crude vinegar, or it may look like jelly. If it looks like jelly, it is very likely that when you put it in your vinegar crock, it will sink and thereby go dormant for lack of oxygen. To prevent this, a piece of thin wood about the size of a playing card is floated on top of the wine or beer, and the vinegar mother is placed on it. This piece of wood is usually made of oak and is called a vinegar raft.

Vinegar mothers are available as white wine, red wine, beer/malt, and cider. All of them have the same acetic acid bacteria, and the only difference is the carrier. In small batches of vinegar—say less than a gallon—the carrier makes a difference in the flavor, but in larger batches of vinegar the carrier doesn't matter.

Some strains of acetic acid bacteria, such as gluconobacter oxydans, will go dormant once all of the ethyl alcohol has been consumed. But the *Acetobacter aceti* that

---

24 Acetobacter aceti, gluconobacter oxydans, and acetobacter pasteurianus predominate.

25 A. Joyeux, S. Lafon-Lafourcade, and P. Ribéreau-Gayon (1984), "Evolution of Acetic Acid Bacteria During Fermentation and Storage of Wine," *Appl Environ Microbiol.* 1984 July; 48(1): 153–156

you'll be using does not go dormant once all of the ethyl alcohol is used. Instead, it starts consuming the acetic acid that it produced, with the end result being just carbon dioxide. So vinegar conversions using a commercial vinegar mother must be arrested once the conversion has completed or you'll end up with no vinegar at all.

The conversion process can be stopped in two ways. For purposes of aging the vinegar, it can be placed in a canning jar with a tight-fitting lid that excludes oxygen. This leaves the vinegar alive, but dormant. For purposes of long-term storage or use in an environment where oxygen might be admitted, the vinegar is pasteurized. Vinegar is pasteurized by heating it to 150 degrees for 30 minutes with the lid adjusted as for canning to prevent evaporation. Once it has been pasteurized, it can be stored in any clean container for a nearly indefinite period of time.

Ethanol to Acetic Acid Conversion

If you are using commercial beer or wine to make vinegar, the amount of alcohol (by volume) is listed on the label. If you are using your own, you should have a good idea how much alcohol is in your beer or wine from the hydrometer readings you recorded.

The chemical equation for the conversion of ethanol to vinegar is:

$$C_2H_5OH + O_2 \rightarrow CH_3COOH + H_2O$$

So ethanol plus oxygen gets converted to acetic acid plus water. Looking at the equation, each molecule of ethanol is converted into one molecule of acetic acid. The molecular weight of ethanol is 46.07 and its density is .789 g/cm^3. The molecular weight of acetic acid is 60.5 and its density is 1.049 g/cm^3.

This means that every gram of alcohol will result in 60.5/46.07 or 1.313 grams of vinegar. A gram of alcohol will occupy 1/.789 or 1.27 cm^3. Alcohol percentages are done by volume, but vinegar percentages are done by weight. We can get a good idea of the conversion factor, that is, how much acetic acid a given amount of ethanol will create, by doing the math for a hypothetical 10% wine.

If I have a liter of 10% wine, that liter contains 100 ml of alcohol. 100 ml of alcohol has a mass of 100 cm^3/1.27 cm^3 or 78.7 grams. The mass of the vinegar produced will be 78.7 * 1.313 or 103 grams.

Therefore, a 10% alcohol by volume wine will create a 10.3% by weight vinegar. So in essence the percentages are identical. Knowing this fact will allow us to dilute the beer or wine we are adding to the vinegar mother to produce a vinegar of known strength. We'd still test it just to be sure, of course. But this allows us to make our vinegar very precisely.

How to Safely Use Homemade Vinegar in Canning

All canning books tell you to never use homemade vinegar in canning. That's because pickling recipes rely upon the vinegar having a certain strength of 5%, and if you use vinegar of a lesser strength you could wind up with botulism-tainted food that could kill you. So if you don't know for sure that the strength of your vinegar is 5% or greater, you can't use it safely. Of course, if the vinegar is substantially stronger than 5% you could wind up with pickled foods that are a lot more acidic than you'd like. You can always dilute it if it is too strong.

The solution to this problem is to figure out how much acidity is in the vinegar. You can do this easily by using the ingredients in a standard acid testing kit available from all wine-making suppliers, a pH meter, and a slight change in procedure. I specify using a pH meter rather than the phenolphthalein indicator because phenolphthalein turns pink when the endpoint is reached, and such a color change may be difficult to discern in vinegar of certain colors. A pH meter won't trick your eye.

Equipment

> 1 50ml beaker
> 1 10ml syringe (no needle needed)
> 0.2N Sodium Hydroxide solution
> Distilled water
> pH meter

Rinsing the syringe using distilled water after each use, put 2ml of the vinegar to be tested and 20ml of distilled water in the 50ml beaker. Fill the syringe with 0.2N sodium hydroxide solution to exactly the 10ml mark. Initially, add 1ml of sodium hydroxide to the beaker each time, swirl, then test with the pH meter. As the pH approaches its endpoint of 8.3, use lesser quantities. Repeat until the solution has a pH of 8.3.

The amount of acid in your vinegar is given by the following equation:

Percentage Acetic Acid = 0.6 x (10−reading of syringe at endpoint)

12

Vinegar Making Techniques

Making vinegar is easier than making wine or beer and requires minimal equipment or ingredients. Other than a vinegar crock and the wine or beer you'll be using, you can get everything else you need for under $30. Here are the items you'll need:

Vinegar Crock

Vinegar can theoretically be made in any sort of container. Traditionally, it is made in oak barrels called vinegar casks or in ceramic urns known as vinegar crocks.

There are three important features in a container used to make vinegar. The container should have a mouth wide enough that you can insert your vinegar raft and preferably your whole hand. It should have a tap, spout, or spigot near the bottom, but far enough from the bottom that it doesn't pick up sediment. Finally, it should be made of a material that will not react with the vinegar. Vinegar is a dilute acid, so it will react with most metals given time.

Given these features, you are not constrained to only use products officially sold as vinegar crocks. Anything officially sold as a vinegar crock will quite frankly be seriously over-priced. I looked on the Internet recently and found many of them priced at nearly $100!

I use two containers to make vinegar. One is a miniature ceramic water crock that holds a half gallon. It costs $24. The other is a one-gallon plastic beverage dispenser I picked up at a department store for $4. Both of these containers have the essential features, including the spigot. Normal ceramic water crocks hold 2½ gallons, an amount which may far exceed the amount of vinegar you plan to make. That's why I got a miniature ½ gallon crock.

You could go all out and get an oak vinegar cask, but that will set you back at least $80. If you want your vinegar to be oak-aged, just add oak cubes to the sealed pint or quart jar that you are using to age your vinegar.

Cheesecloth and Rubber Bands

These items are used over the mouth of your vinegar crock to allow oxygen to enter but keep fruit flies and other critters out. Not all cheesecloth is created equal. The material that is sold as "cheesecloth" at the supermarket is not suitable for making cheese, and even doubled or tripled it won't keep fruit flies out of your vinegar.

Unless you have a good gourmet shop nearby that sells real cheesecloth, you may have to order it from a supplier of cheese-making supplies over the Internet. It is a bit expensive when you include shipping, so I recommend saving on shipping by ordering a couple of packages. They won't go to waste because you'll need the cheesecloth for making cheese in the next chapter.

The size of the needed rubber bands will be different depending upon the size of the mouth of your vinegar crock. The only caution

❯❯ Use a doubled piece of high-quality fine cheesecloth to cover your vinegar urn. Otherwise, fruit flies will get into your vinegar.

worth mentioning is that light and vinegar fumes will degrade the rubber, so check the rubber bands weekly and replace them if you see signs of deterioration. Otherwise you'll look at your crock one day and find more flies in it than vinegar.

Miscellaneous Supplies

A vinegar raft is a small thin piece of oak that floats on top of your vinegar. Its purpose is to keep the vinegar mother from sinking

❷ The materials for oak-aging vinegar are simple and inexpensive.

because if the vinegar mother sinks, it will stop making vinegar. These are available in vinegar kits or individually from many Internet sites. Just type "vinegar raft" into a search engine.

Some people prefer the taste of vinegar that has been aged in oak, or the astringency contributed by the tannins leached from the oak. Oak barrels are expensive and time-intensive to maintain. An alternative is adding oak chips or oak cubes to the vinegar. Add a quarter cup per gallon, enclosed in a tied spice bag for easy removal later. The chips or cubes are added during the aging process and left in the vinegar for four to six weeks. For these purposes, you don't want to use oak from your building supply store. Instead, order it from a winemaking supplier. Winemaking suppliers can offer a range of oaks with different taste characteristics that you know aren't contaminated with anything nasty.

Canning jars are a good choice for aging and storing vinegars. They seal tightly, which will cause the vinegar mother to go dormant during aging, and they can be used repeatedly which makes them a good bargain.

One other thing you may find helpful is a funnel that you have attached to a piece of plastic hose such as the hose used for racking wine. As vinegar is being made, you need to add more beer or wine. The easy way to do this without risk of disturbing the vinegar mother is to insert the hose into the liquid in the vinegar crock, and add the liquid through the funnel.

A candy thermometer will be needed for pasteurizing vinegar, unless you plan to can it using a boiling water bath canner for long-term storage.

Consolidated Equipment and Ingredient List

- Vinegar crock
- Vinegar mother
- Cheesecloth
- Rubber bands
- Vinegar raft
- Canning jars
- Candy thermometer
- Oak chips or cubes (optional)

Making Your Vinegar

The first thing to do is pre-dilute your wine or beer if needed. At levels higher than 7% alcohol it might inhibit the AAB. You can always make it less concentrated, down to 3%, for purely culinary use or if your beer only has that much alcohol, and it isn't unusual for sherry vinegars to be as high as 7%. In general, I recommend diluting to 5.5% so the vinegar can be used with greater versatility. Always dilute with clean, non-chlorinated water. I use bottled water for this purpose.

So . . . how much water do you add to your beer or wine to get a certain percentage of alcohol? Start by dividing the current concentration in percent by the desired concentration in percent to get C. So if I have some 10% wine and I want 5.5%, I divide 10 by 5.5 to get 1.82. Next, multiply the volume of your wine (say 500 ml in a standard wine bottle) by C to get the total diluted volume: 500ml × 1.82 = 910. Finally, subtract the volume of wine from the total volume to get the volume of water you need to add. 910 ml – 500 ml = 410 ml.

This also works with beer. Say I have some beer that is 6% alcohol and I want to dilute it to 5.5%. The standard beer bottle is 12 ounces. So C = 6%/5.5% = 1.09. Multiply 12 oz x 1.09 = 13. Finally, 13 – 12 = 1, so I would add one ounce of water.

The quantity of diluted wine or beer that you use is important because it takes a while for the vinegar mother to work, and in the meantime the underlying beer or wine is vulnerable to outside infection. You want to limit the amount you put in the crock to no more than triple the volume of the vinegar mother, which is eight ounces. So your initial ingredients of the vinegar crock will be 24 ounces of beer or wine diluted as needed and eight ounces of vinegar mother for a total of 32 ounces.

Making Vinegar, Step by Step

1. **Clean your vinegar** crock thoroughly and sanitize it using sulfite solution. (See the chapters on wine for how to make sulfite solution.)

2. Check the capacity of the container of vinegar mother you ordered. Usually it is eight ounces.

3. Add diluted wine or beer to the vinegar crock. The amount added should be twice the volume of the vinegar mother. So if you have eight ounces of vinegar mother, put 24 ounces of wine or beer in your crock. The alcohol percentage cannot exceed 7%.

4. Open your vinegar mother. If it is gelatinous, place your vinegar raft on top of the water/wine solution in the vinegar crock.

5. Add the vinegar mother. If it is all liquid, just gently pour it into the crock. If it is gelatinous, add it on top of the vinegar raft.

6. Cover the mouth of the container with cheesecloth and hold it in place with a rubber band.

7. Set the container in a dark place or at least someplace well out of the sun. The ideal temperature range is 80 to 90 degrees, but it will progress fine at 70 to 100.

8. Depending on temperature and other factors, the complete conversion of wine to vinegar can take anywhere from six weeks to three months. Check your vinegar weekly by sniffing it through the cheesecloth. It should smell like vinegar is forming.

9. To increase the volume of the vinegar being made, you can add more diluted wine or beer starting at the fourth week and every fourth week thereafter. Add by using a sanitized funnel and tubing.

10. Six weeks after the final addition of wine, start tasting small (less than ¼ tsp) samples of the vinegar to see if it is done. It's done when all the alcohol flavor has been replaced with vinegar flavor. Your tongue and nose are amazingly sensitive and able to detect many substances in very low concentrations of parts-per-million. This is as accurate as any easily performed test in determining if the vinegar is done.

11. Once the vinegar is done, it is important to remove it from the vinegar crock because with all the alcohol gone, the vinegar mother will start

❯❯ The mother in this vinegar crock is doing nicely.

consuming the acetic acid, and thereby destroy the vinegar. Take out as much vinegar as you can through the spigot and then start your next batch using the same vinegar mother in that container. As long as your vinegar doesn't become contaminated, you can use the same vinegar mother indefinitely.

Aging Vinegar

Just like wine, vinegar made from wine will mellow with age. Freshly-made vinegar is very sharp with a lot of pointed edges. When it is allowed to age, the compounds within the vinegar combine in various ways that make the vinegar more mellow and to bring out other flavor components.

Even though it is easy to visualize the vinegar mother as sitting on top of the wine, many of its bacteria are spread throughout the vinegar. When you draw off a sample, even if it looks clear, it is filled with acetic acid bacteria. (These bacteria, incidentally, are totally harmless to humans.) Freshly-made vinegar is teeming with life.

When vinegar is aged, it is aged with that life intact. The vinegar is drawn from the crock via the spigot and placed in a container sealed so as to exclude air. This renders the acetic acid bacteria dormant. Vinegar can be kept in a sealed container for an indefinite period of time. In fact, genuine balsamic vinegar is aged for at least twelve years, and often for as long as 25 years. The minimum period of aging I would recommend is six weeks.

Vinegar can be aged in porcelain, glass, impervious plastic, or wooden barrels. A lot of the better traditionally-made vinegars feature oak aging. The oak aging serves to impart an astringent principle to the vinegar in the form of tannin. Tannin is not just one substance. The term "tannin" refers to literally dozens if not hundreds of related compounds formed around either a gallic acid or a flavone core. Tannins have in common not only their astringency, but also their ability to bind and precipitate proteins. This means that tannins introduced into vinegar will scavenge stray proteins left over from fermentation by combining with them to form an insoluble substance that will sink to the bottom of the container.

So over time, an initially high level of tannins is reduced and a number of protein- or amino acid-based substances are removed. This serves to alter the flavor in more ways than merely introducing astringency. In fact, the addition of tannin, through its ability to remove other substances, can paradoxically decrease the astringency of vinegar over a period of aging by removing other substances. Tannins also combine with metals in a process known as chelation. Chelation forms soluble compounds that include the metal but render it unavailable to combine with other substances. This likewise affects the flavor.

You can use oak in the aging of your vinegar by placing a quarter cup of the cubes or chips in a tied spice bag in your aging vinegar. Leave it in the container for six weeks, and then remove it using sterilized tongs and re-seal the container. The rest of the vinegar's aging will continue to be affected by the tannins imparted by the oak.

Keeping Vinegar

Eventually, the aging process ends and the vinegar is ready for storage. The next step is to filter and pasteurize. Perfectly adequate filtration is achieved by pouring the vinegar from the jar in which it is aging through a funnel lined with a coffee filter into a clean canning jar.

Fill the jar with vinegar to within a quarter inch of the top, and install the two-piece canning lid. Process for 10 minutes in a boiling water bath or steam canner and pasteurization is complete.

Making Herbal Vinegars

I'll confess that I have never purchased an herbal vinegar. Anytime I have seen herbal vinegar, it is usually in some sort of craft shop. The vinegar is in an ornate bottle with a sprig of some herb and has a fancy label. It also has an obscenely fancy price. The price seems crazy to me because I'm pretty certain that the vinegar they used was $1/gallon commodity vinegar and the sprig of herb cost about a penny, and the cost is $12 for six ounces. No thanks!

Herbal vinegars can be quite nice, though, and making your own is easy enough. You can make it using commodity vinegar from the supermarket or your own hand-crafted vinegar. I don't recommend using cider vinegars for herbs.

As you may know from an earlier book,[26] growing your own herbs is easy. The hard part for beginners is choosing which herbs to use (and how much). To help you get started, I suggest the following single herbs: borage, thyme, rosemary, dill, basil, tarragon, and oregano.

I recommend making your herbal vinegars from fresh herbs when possible. Using fresh herbs, I recommend ½ ounce of fresh herb per cup (eight ounces) of vinegar as a starting proportion. Because the vinegar is a preservative, the herbs won't rot. When using dried herbs, use two tablespoons of dried herb per cup of vinegar.

The procedure is straightforward. Add the cleaned herbs to the container that will hold the herbal vinegar. Heat up the vinegar to a simmer (NOT a boil!), and

---

26 *Maximizing Your Mini Farm*

then pour the vinegar into the container holding the herb. Seal the container. Allow the flavors to meld for three or four weeks to develop the full flavor before using.

If you want something really impressive for making an oil and vinegar dressing for salad, I would suggest making vinegar from pear wine, and then using the pear vinegar to make a borage herbal vinegar.

Oil and Vinegar Dressing

Ingredients:

11 ounces	Virgin olive oil
5 ounces	Hand-crafted wine vinegar
2 ounces	Water
1 Tbsp	Pulverized dehydrated sweet red pepper
1 Tbsp	Pulverized dehydrated onion
1 tsp	Sea salt
1 tsp	Garlic powder
1 tsp	Dried oregano
1 tsp	Dried basil
⅛ tsp	Xanthan gum OR ½ tsp dried powdered purslane or okra (optional)

Procedure

Add ¼ cup of water and ½ cup plus 2 Tbsp vinegar to your container. Add the remaining solid ingredients except for the xanthan gum/purslane. Shake and allow to sit for a few minutes. Add the xanthan gum/purslane and shake thoroughly. Add 1¼ cup plus 2 Tbsp of olive oil. Shake thoroughly.

The purpose of the xanthan gum or purslane in this recipe is to keep the mixture from separating too quickly for practical use because oil/vinegar and oil/water don't normally mix. The xanthan gum or purslane helps to keep it in suspension. If you use xanthan gum, don't use more than the recommended amount or you'll end up with a jelly-like substance rather than dressing.

PART V
Cheese Making

13

Cheese: Ingredients and Equipment

Protein is an essential part of the human diet. Though vegetable sources can provide protein, in most cases the protein lacks crucial amino acids. The most readily available complete proteins are meats, eggs, and dairy; the latter two are the least expensive. Continuing the theme of preserving nutritive content through fermentation, we arrive at cheese. Milk contains a lot of complete protein, but it is also highly perishable.

In the ages before refrigeration was reliably available, one of the few ways to make the nutritional value of milk last longer while also making it quite portable was turning it into cheese. Hard cheeses in particular, if waxed, can last for years.

Another advantage of cheese is that many hard cheeses lack lactose. Lactose is a sugar in milk that many folks (including myself!) cannot digest. As a result, if they consume most milk products they will suffer severe gastrointestinal distress—sometimes for days. When the whey and curd are separated in the first phases of making cheese, 94% of the lactose stays in the whey. Most aged cheeses lack lactose and as a result provide lactose-intolerant people with a delicious way of obtaining the nutritional benefits of milk.

Cheese also has its own health benefits. It is rich in cancer-preventing conjugated linoleic acid and sphingolipids, fights tooth decay, and helps maintain bone strength.

Like beer making, cheese making is both art and science. If anything, there is even more art to making cheese because it requires practice to master the various steps. So this chapter is enough to get you started, but you'll likely want to branch out once you've mastered the techniques covered here.

What is Cheese?

Cheese is the coagulated fat and protein from the milk of domesticated dairy animals. The fats and proteins of milk are coagulated in various ways for the manufacture of different types of cheese. In some cases, a bacterial culture is added. The bacterial culture consumes lactose to make lactic acid; this lactic acid causes the coagulation.

In other cases, rennet is added. Rennet is a complex mixture of enzymes that likewise coagulates milk. In yet other cases, an acid such as citric acid, tartaric acid, or even vinegar is used to cause coagulation. Though the products of these various methods of coagulation are markedly different, they are all cheese because they have in common the coagulation of milk.

Milk: Where it all Begins

In the United States, cows are the usual source for milk; goats are utilized to a lesser extent. In other countries, the milk of bison, buffalo, sheep, horses, yaks, and other animals are also used. The nature of the milk of different species varies appreciably and this is reflected in the character of the cheese produced. Theoretically, you could make cheese using the milk of any mammal; I wouldn't attempt this until you get good at making cheese from well-characterized herbivores such as cows and goats. Not only that, trying to milk a tiger or a bear is probably more dangerous than warranted.

» Most organic milk is ultra-pasteurized, making it unsuitable for cheese.

Likewise, the components of the milk will vary between different breeds of dairy cattle. Even the milk of a particular cow will vary with season and diet. Probably the most striking example of this was in the cream cheese my grandmother would make from cows that had been eating wild onions. The smell and taste of the wild onions was transferred to the milk and hence to the cheese. In the case of cream cheese, the results were delicious!

It is important to know that though pasteurized milk is fine for making cheese, the ultra-pasteurized milk that you find in the store is unsuitable. This is unfortunate, because it is the organic brands that tend to be ultra-pasteurized. Ultra-pasteurization is used to extend the shelf-life of expensive milk that doesn't sell very quickly. Unfortunately, that process damages the protein in milk so extensively that it is unsuitable for making cheese.

Milk from other animals can certainly be made into cheese, but doing so would require changes in timing, temperature, quantities of ingredients, and so forth that are simply too extensive to be treated in a single chapter.

So we are going to use pasteurized, homogenized cow's milk from the grocery store as the learning medium for your first forays into cheese making. After you have mastered these skills, you can branch out from there. You can find specific types of milk suitable for your needs by finding a local dairy at www.smalldairy.com.

About Raw Milk

Cheese connoisseurs insist that the best cheeses are made from raw milk that has been neither pasteurized nor homogenized. The trouble is that raw milk is not readily available and quite often there are legal impediments to buying it directly from farmers. The basis for these legal impediments is widespread recognition of the likelihood of the presence of pathogens in raw milk.

In former times the largest risks of raw milk were brucellosis and tuberculosis; today the risks are e. coli, salmonella, and listeria. Testing of vats of milk in modern times shows that even from healthy cows, anywhere from 0.87% to 12.6% of raw milk harbors dangerous pathogens.[27] How do healthy cows give pathogen-infested milk? They don't. Inadequate sanitation and cleaning of equipment introduces fecal bacteria into the milk. The reason pasteurization became a requirement in the first place was that farmers were actively falsifying their records so that tuberculosis-infected cows wouldn't have to be removed from milk production.[28]

The reason it continues to be required is because human nature hasn't changed, and maintaining sanitation on an industrial scale of a biological product created by an animal that excretes feces requires extreme levels of conscientiousness that cannot be guaranteed. In essence, because the healthiness of cows and their milk can be tested to assure a safe product without pasteurization, it is possible to sell perfectly healthy raw milk. But pasteurization is required anyway to compensate for the existence of lazy or dishonest people that will prioritize the production of a single infected cow over the health and well-being of their customers. I'm quite sure most people would do the right thing, but in an industrial system where the outputs of various farms are mixed together, it only requires one feces-contaminated vat to sicken thousands of people.

Obviously, raw milk that does not contain pathogens can be made. Humans have consumed raw milk for thousands of years before pasteurization was invented. Such milk was collected at home by the end users, so there was a direct correlation between shoddiness and adverse consequences that would result from collecting milk in a bucket that wasn't clean. The milk was used immediately rather than transported thousands of miles, so any pathogens present had less opportunity to multiply to dangerous or infective levels. It is therefore possible to obtain raw milk that will not make you sick, provided it is supplied by an honest and conscientious farmer.

How to determine if someone is honest and conscientious, I can't say. If I could write a book describing a sure-fire technique of that sort, personnel managers across the world would rejoice. In the absence of that, I would instead look at the idea of mutual self-interest. If a farmer were to sell you raw milk that made you sick, your family could sue him into oblivion. So it is in his best interest, if he sells raw milk at all, to make sure it is pristine. Many such farmers use small-scale low-

---

27 Position Statement on Raw Milk Sales and Consumption, Cornell University Food Science Department

28 "Not on My Farm!: Resistance to Bovine Tuberculosis Eradication in the United States," Alan L. Olmstead and Paul W. Rhode, January 2005, *The Journal of Economic History* (2007), 67 : 768-809 Cambridge University Press, Copyright © 2007 The Economic History Association, doi:10.1017/S0022050707000307

temperature vat pasteurization just to be sure, and this process is less damaging to the milk proteins than standard pasteurization processes.

One other layer of protection is to only use raw milk to make hard cheeses that are aged for longer than two months. The process of cheese-making, when combined with the conditions of aging in cheese, serve to eliminate potential pathogens and render the cheese safe. This only applies to aged hard cheeses! Soft cheeses and those eaten less than two months from manufacture should be considered as risky as raw milk, and I personally avoid making cheese from raw milk, but that's an individual choice.

If you use raw milk in cheese-making, there are only two procedural changes you'll need to adopt. The first is that you can avoid using calcium chloride (described later), and the other is that when heating the milk, especially for thermophilic cheeses, you will need to top-stir the milk. Top stirring is just slowly dragging a utensil across the top quarter-inch of milk in order to keep the milk fats from separating out.

To find raw milk, I recommend the following Internet resources:

- A Campaign for Real Milk: www.realmilk.com
- The Weston A. Price Foundation: www.westonaprice.org
- Farm-to-Consumer Legal Defense Fund: www.farmtoconsumer.org

Categories of Cheese

Cheese can be categorized in various ways depending upon the substances from which it is made, its appearance or consistency, whether it is aged or eaten fresh, and the procedures used to produce it. For our purposes, we will use fresh and aged cheeses as categories, as well as soft and hard cheeses, since these categories have the greatest differentiation.

Equipment

When it comes to the equipment needed to make cheese, quality matters. The good news is that most of this equipment is a once-in-a-lifetime purchase.

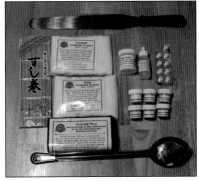

>> Quality ingredients and equipment will contribute to a quality product.

You will likely end up ordering most of these items over the Internet because you may have difficulty finding them locally.

Measuring Cups and Spoons

You want both a large (2+ cup) and small (1 cup) Pyrex™ glass liquid measuring cups. You will also need measuring spoons, but not the ordinary cheap ones you get at the dollar store. You want high-quality stainless steel measuring spoons that measure in ⅟₃₂, ⅟₁₆, ⅛, and ¼ teaspoon increments, as well as the traditional sizes.

I have noted by comparing volumes to my laboratory standards that cheap measuring spoons are often undersized or over-sized. This is not a critical matter when making a cake; when making cheese it can spell the difference between success and failure.

Large Double Boiler

With batches of cheese starting with a gallon of milk or less and that use a mesophilic starter culture (more on starter cultures later), you can get by with a standard large pot that you set in a sink of hot water. For batches of cheese requiring more than a gallon of milk or using a thermophilic starter culture, you will need a double-boiler. In cheese-making, this double-boiler is also called a "cheese pot." For very small batches of cheese starting with a quart of milk, you can improvise by setting a smaller pot into a larger one as long as the handles on the smaller pot will sit on the lip of the larger pot so the smaller one is surrounded by water.

Again, depending on the size of your largest intended batch of cheese, you may be able to use a double-boiler as small as eight quarts. But because it takes a large amount of milk to make enough curd to yield very much hard cheese after pressing, you won't go wrong with a boiler as large as 20 quarts. No matter what size you use, make sure it is stainless steel because acidified milk will leach aluminum or iron into your curd and impart metallic flavors.

If you don't already have a double-boiler, this is probably the most expensive item you'll need to get. Searching the Internet, I found prices ranging from $88 to $130 for a 20-quart model. It won't come cheaply, but you'll be thankful that you got it. You can use it for batches of cheese starting with anywhere from one gallon to four gallons of milk, and its configuration will help to hold temperatures steady while preventing scorching.

Colander

You'll need a large eight-quart colander that will fit into the cheese pot with the handles resting on the edges of the pot. You'll use this to separate the curds from the whey, with the whey going back into the pot.

Special Utensils

You need a large stainless steel slotted spoon, a stainless steel skimming ladle, and a stainless steel curd knife. This latter utensil is specialized so you will probably have to get it via an Internet source.

Cheesecloth

You want high-quality coarse (20 thread count) and fine (60 thread count) cheesecloth. The fine cheesecloth is used for making soft cheese such as cream cheese; the coarse cheesecloth is used to hold harder cheeses during the pressing or curing process.

Cheesecloth is packaged in two-yard increments, so you get a piece that is three feet wide and six feet long. Cut off pieces as needed with good scissors. Before use, cheesecloth must be sterilized. Put it in a pan of water, boil for five minutes and then dump the cheesecloth and water into a colander in the sink. Cheesecloth can be re-used. Rinse it under cool running water, work a few drops of dish liquid into it, rinse it thoroughly, and boil it for five minutes. After boiling, hang it up to dry, then store the dried cheesecloth in an airtight bag. Don't forget to sterilize it before using it again.

Bamboo Sushi Mats

These allow good air circulation for cheese that is either draining or aging, and is essential in making hard cheeses. Luckily, they are inexpensive at $4 each or less. They can't be sanitized and should be discarded after use.

Cheese Wax or a Vacuum Sealer

Cheese wax is used to protect the cheese from air while it ages. This is a special kind of wax that melts at a low enough temperature that it won't hurt the

cheese when you brush it on. Don't try to substitute canning wax for this! Another alternative is to use a vacuum sealer to seal the cheese in an airtight bag from which all air has been evacuated. That's what I do because it is more convenient than waxing.

Cheese Press and Mold

A cheese press is used to knit the curds together into a solid mass while expelling excess whey. There are a variety of designs of varying expense and complexity. A search on the Internet will even reveal many free design plans for making your own.

For most of the batches of cheese I've made, I have used a simple plastic press and mold that only cost $21. The downside is that you have to use external weights with it. Still, you can't beat it for the price and ease of use. Recently, I have acquired a stainless steel press made by Wood Lab that works very well.

Instant-read Digital Thermometer

Temperatures are critical when making cheese. Some types of cheese require gradually raising the temperature or holding at a certain temperature for a specified time. The best thermometer for such purposes is one that gives you an instant and accurate reading. A good digital thermometer is not expensive.

I have a "Norpro electronic digital read thermometer/timer" that cost $16 and a hand-held Hamilton Beach instant-read digital thermometer. Both cost under $20, have stainless steel probes that are easy to sterilize, and can be found at cookware stores.

Dedicated Small Refrigerator

Traditionally, many styles of cheese were quite literally aged in caves. Caves maintain a constant temperature and humidity throughout the year. Most of us don't have access to a suitable cave, and we don't have an area in the house that will reliably maintain a certain temperature for months on end.

If you decide to make cheeses requiring aging, you will find a dedicated refrigerator indispensable. A second-hand dormitory-sized refrigerator and an external thermometer set up to turn it on and off as needed will work perfectly for such an endeavor. A refrigerator dedicated to cheese-making is called a "cheese cave."

Ingredients

Not all of these ingredients are needed for all cheeses, but you'll want them on-hand. Some of these you may already have from your excursions into wine, beer, and vinegar making.

Vinegar, lemons, and tartaric acid

These common acids are used to make soft cheeses via the direct acidification method. In this method, the milk is heated to a certain temperature, a measured amount of acid is added and stirred into the milk, and then the milk clots after a period of time. This clotted milk is poured into a colander lined with cheesecloth; the cheesecloth is tied into a bag. The bag is hung in a warm place for the whey to drain out of the soft curds. These are among the easiest cheeses to make, and they work especially well as dips and spreads.

Calcium chloride, 30% solution

When milk is pasteurized, the calcium ion balance is upset in the milk, which can impede proper curd formation. A small amount of calcium chloride solution diluted further in distilled water and mixed into the milk can correct this imbalance.

You can order food-grade calcium chloride and make the solution yourself (percentages are by weight!), or you can order the pre-mixed solution from various Internet stores specializing in cheese making supplies.

Calcium chloride, incidentally, is also an ingredient in some ice melting pellets used to melt the ice on sidewalks and driveways. This is a very crude product that isn't suitable for human consumption, so make sure you get food grade calcium chloride.

Flaked or canning salt

Salt is used as a flavor enhancer, a bacteriostatic preservative, a modulator for enzyme action, and it helps expel water from cheese curds through osmotic pressure. Special "flaked" cheese salt is available, but canning salt or Celtic sea salt will do as well.

The important thing is to avoid the ordinary salts in the grocery store because not only do many of them contain iodine, they often contain anti-caking agents and other chemicals that could interfere with cheese-making. So anything you use should be purely salt.

Starter Culture

You can buy starter culture in packets from a supplier, or you can make your own from buttermilk and yogurt. Starter cultures are either mesophilic (meaning "medium heat-loving") or thermophilic (meaning "high heat-loving"). Starter culture is an inoculant containing a mix of bacteria that eat the lactose in milk and excrete lactic acid. The first purpose of these bacteria is to lower the pH of the milk in order to encourage curd formation. The second purpose is the continuing development of flavor characteristics during the making and aging of the cheese. The nature of the starter culture strongly influences the flavor of the cheese.

Mesophilic starter cultures work best at room temperature—around 72 degrees. They usually contain at least *Streptococcus lactis*, and many also contain *Streptococcus lactis* var. *cremoris* along with other lactic acid bacteria such as *L. delbrueckii* subsp. *lactis*, *L. lactis* subsp. *lactis* biovar *diacetylactis*, and *Leuconostoc mesenteroides* subsp. *cremoris*.

Streptococcus lactis is used to make cultured buttermilk; therefore fresh buttermilk with active live cultures can be used to make a mesophilic starter culture for cheese-making. Cheeses that begin with a mesophilic starter include farmhouse cheddar, edam, stilton, and Monterey Jack, among others.

Thermophilic starter cultures work best at temperatures above 80 degrees and below 130 degrees. A specific recipe will dictate the best temperature within this range for the particular cheese being produced, but the culture works best at 110 degrees. Exceeding 130 degrees may kill a thermophilic culture. This culture may like heat, but it doesn't want to be scalded or boiled. Thermophilic starters are used to create Swiss and Parmesan cheeses among others. *Streptococcus thermophilus* is a common bacteria in thermophilic starter cultures, but *Lactobacillus delbrueckii* subsp. *bulgaricus*, *L. delbrueckii* subsp. *lactis*, *L. casei*, and *L. plantarum* are all used.

Yogurt is made with thermophilic bacteria. One prominent brand of organic yogurt uses six live cultures that include *Streptococcus thermophilus, Lactoba-*

cillus delbrueckii subsp. *bulgaricus*, and *L. casei*. This means that plain yogurt can be used to make more yogurt, and it can also be used to make a thermophilic starter culture for cheese.

If you opt to buy starter cultures from a cheese-making supply store instead of making your own, there are only two important things you need to know: You want the sort of culture called a "direct vat" culture, and you should put it in the coldest part of your freezer the very second you get it. Keep it in the freezer until ready for use.

Rennet

Rennet is an enzyme that was originally derived from the stomachs of suckling animals. It is a proteolytic enzyme that breaks protein bonds in such a way as to turn liquid milk into solid curds. All infant mammals produce rennet. This turns milk into a solid form that stays in their digestive tract longer. That's why when a baby spits up milk, it has mysteriously turned into a clumpy solid. Babies of all mammals have miniature internal cheese factories.

In practice, animal rennet is a byproduct of veal production. Animal rennet of this sort is extremely perishable and has to be kept refrigerated. It's also pretty expensive.

Rennet can also be made from certain fungi and plants. The sort made from plants has to be made fresh on the spot, which may not be feasible during winter or if you can't find the plants. For our purposes I am recommending vegetable rennet, which is actually made from fungi. It is inexpensive and if you put it in the freezer it will stay good for about six months. It comes in tablets that can be divided into halves and quarters; this must be done carefully as it has a tendency to disintegrate.

Rennet is an extremely powerful enzyme. Tiny quantities will clot gallons of milk. When adding rennet, dissolve the required amount into a quarter cup of distilled water over a period of 20 minutes, then sprinkle it over the surface of the milk. Mix it into the milk using up-down and back-and-forth motions rather than swirling because swirling doesn't mix as efficiently. It's important that rennet be mixed efficiently because otherwise the curd it forms will be of uneven consistency.

How to Make Rennet from Nettles

In a pinch, you might need to make your own rennet from nettles. This rennet works, but it doesn't give as clean a break or as solid a curd.

Put a pound of stinging nettle tips in a large pot and cover with water. Bring to a light boil and boil until the volume has been reduced by half. Filter through cheesecloth into a clean container. You can keep this in the refrigerator for up to two weeks. You use one cup per gallon of milk to be curdled.

Other Cultures and Enzymes

As your cheese-making expertise increases, you'll want to try to make specific types of cheese. Toward that end, you will need different cultures and enzymes.

Lipase is an enzyme that splits milkfat into free fatty acids. It develops a characteristic picante flavor in the manufacturing of feta, blue, mozzarella, and provolone cheeses. Like rennet, it is extremely powerful. Unless a recipe directs otherwise, use between $\frac{1}{16}$ and $\frac{1}{8}$ tsp of the powder per gallon of milk. Dissolve the powder in a half cup of cool water for 30 minutes prior to use. Lipase is added immediately before rennet by sprinkling it on top of the milk and mixing it in using an up-down and back-and-forth motion.

Propionic Shermanii culture is used to create the characteristic holes and flavor of Swiss cheeses. As it ferments, it creates carbon dioxide that expands to create the holes. This is added to thermophilic starter culture at the rate of $\frac{1}{16}$ tsp per gallon of milk.

Not all mesophilic or thermophilic starter cultures are created equal. The specific varieties of bacteria make a difference in the ultimate flavor of your cheese. As you learn more about cheese, you will want to try other starter cultures.

14

Practical Cheese Making Techniques

In this chapter, I am going to progress from the easiest and least time-consuming techniques to the more involved, using a few examples. By mixing, matching, and varying these techniques you can make a wide array of interesting cheeses. Buttermilk and yogurt are an ideal starting place because both can be used to make other cheeses while saving money on starter cultures.

How to Have a Lifetime Supply of Buttermilk and Mesophilic Cheese Starter

I have always loved cultured buttermilk. Its thick consistency with sweet-tartness is irresistibly delicious, and it makes wonderful pancakes as well! Buttermilk costs 70% more than regular milk, so if you like it, you can save money by making your own.

Start with cultured buttermilk from the store that uses live cultures. You can make any amount of buttermilk you'd like from this by re-culturing. To re-culture, put the amount of milk you would like to turn into buttermilk into a stainless steel container. Either use a double-boiler or put the container of milk into a sink of hot water, and raise the temperature to 86 degrees.

Hold at 86 degrees for ten minutes, then add ¾ cup of buttermilk per quart of milk. (1½ cups of buttermilk for a half-gallon and 3 cups of buttermilk for a gallon.) Remove the milk from the heat, cover with cheesecloth to keep out bugs but allow oxygen, and allow it to sit at room temperature undisturbed for twelve hours.

That's it. Really. If you refrigerate it after the twelve hours are up, it will keep in the refrigerator for up to two weeks. Anytime you want more buttermilk, just repeat this procedure using a bit of the buttermilk you already made and you can have buttermilk forever unless your supply becomes contaminated.

Anytime a cheese recipe calls for "mesophilic starter" you can use your buttermilk at the rate of four ounces of buttermilk per one gallon of milk that you'll be turning into cheese. It is possible to freeze buttermilk for use later to make cheese, but I don't recommend that as the viability of the culture becomes spotty. I recommend using only unfrozen buttermilk to make cheese.

How to Have a Lifetime Supply of Yogurt and Thermophilic Starter Culture

Yogurt is a bit more difficult to make than buttermilk because it requires the yogurt-in-progress to be held at a higher temperature for a long time. A yogurt-making machine can help, or make the yogurt on a weekend. If your family uses a lot of yogurt, it may be worthwhile to purchase a yogurt machine for less than $100. Yogurt costs anywhere from 300% to 400% more than milk, so if you eat a lot of yogurt you can save a lot of money by making your own.

You can make yogurt successfully from plain yogurt from the store, or you can buy a starter culture for the specific type of yogurt you wish to make. Viili culture produces a thick but mild yogurt similar to what you you mostly see in stores, whereas Piimä culture makes a thinner, drinkable yogurt. There are many other cultures available, but no matter how you start your first batch, yogurt cultures are serial cultures, meaning that you can continue to propagate them indefinitely simply by using a quantity from the last batch to make the next.

If you decide to use plain yogurt from the store to make more yogurt, please read the ingredient label carefully to make sure you are buying a product made only from milk and cultures. There are some yogurt brands whose "plain" yogurt contains adulterants and other ingredients that won't be helpful. Pectin is often used as a thickener and this is okay.

First, heat your milk to 185 degrees in a double boiler while stirring often. This is to kill off competing organisms. Then, remove the milk from the heat and allow it to cool to between 105 and 122 degrees. Once it is between these two temperatures, add either your starter culture according to package directions or ¾ cup of live yogurt per gallon of milk. Pour the mixture into cleaned and sterilized quart canning jars, and adjust the two-piece caps for a seal. Keep the temperature of these containers at 105 to 122 degrees for the next eight hours. The temperature can be maintained by filling the sink with water at 120 degrees, and then adding a bit of boiling water to the water in the sink whenever the temperature drops below 110 degrees. After eight hours, put your jars in the refrigerator where the yogurt will keep for two weeks.

Maintaining this temperature for so long will be difficult, but the bacteria have a better sense of humor than most regulatory agencies, so as long as you keep the temperature above 98 but below 130, your yogurt will still be fine. To maintain this temperature you can use the sink method already mentioned, a mattress heating pad or an electric blanket; be sure you keep an eye on things and check frequently so it doesn't overheat. Or, use your oven if it can maintain temperatures under 120. A slow-cooker with water on the lowest setting may also work by setting the jars in water in the slow-cooker and watching the temperature. The key is to improvise creatively.

The yogurt you create is plain yogurt. You can mix anything with it you'd like—fruit, nuts, granola, sweeteners, etc. If you decide to use it as a thermophilic cheese starter, use four ounces of your fresh plain yogurt per gallon of milk that you will be turning into cheese.

Okay, Let's Make Some Cheese!

There are literally hundreds of types of cheese, all of which require differences in procedure, technique, or ingredients. Rather than try to cover all of it,

I am going to illustrate how to make four representative cheeses that are easily made at home using the ingredients and equipment described. Between these four cheeses, all of the basic techniques will be covered, and you will gain enough experience to experiment and branch out.

I am going to cover a direct acidification soft cheese. Using the same principle, you could make a soft cheese using a different acid. Then, I will demonstrate a soft cheese using a starter culture. Next, I will demonstrate a minimally-aged hard cheese using both starter culture and rennet. Finally, I will describe making an aged cheddar cheese and most importantly the cheddaring technique.

Soft Cheese by Direct Acidification: Queso Blanco

Using a double boiler, raise the temperature of one gallon of milk to 180 degrees while stirring so the milk doesn't precipitate protein. Add ¼ cup of vinegar by slowly dribbling it into the milk while stirring. (You can use distilled vinegar or some of your homemade vinegar. For a different taste, you can use the juice of 3-5 lemons.) Continue to stir for ten to fifteen minutes until the milk is completely clotted. If the milk doesn't clot, add up to four more tablespoons of vinegar while mixing for another ten to fifteen minutes.

⊗ Raising the temperature to 180 degrees before adding the vinegar. Notice the cheesecloth boiling on the right.

Meanwhile, prepare cheesecloth by boiling in a pan of clean water. After boiling, use the cheesecloth to line a colander. Pour the clotted milk into the cheesecloth-lined colander, allowing the liquid to go down the sink. After the cheese has cooled, form the cheesecloth into a bag, and hang it over a bowl until liquid no longer drains out of the bag. (This works best at standard room temperature. If the temperature is too cold, the cheese won't drain well. This process should complete within five to seven hours.)

⊗ The clotted milk draining in the colander.

« I have a hidden hook under my cabinets for hanging cheese to drain.

» This easy cheese is great on bagels or mixed with herbs as a vegetable dip.

Scrape the cheese out of the cheesecloth into a clean, covered container. Add and mix salt, dried herbs such as garlic powder, dill, or basil into the cheese as desired. This is what is called a "fresh" cheese and it should be refrigerated promptly after making. Use within a week to avoid spoilage. Because of all the different things you can mix with this, it is a very versatile cheese that can be used for bagels, dips, and dressings.

Soft Cheese using Yogurt Starter Culture: Farmer's Cheese

Add ½ teaspoon of 30% calcium chloride solution to ¼ cup of water, and mix thoroughly with one gallon of milk in a double boiler. Using the double-boiler, raise the temperature of the gallon of milk to 105 degrees. While the milk is heating, dissolve ¼ of a rennet tablet in ¼ cup of cool non-chlorinated water. Once the milk has reached 105 degrees, keep it there for five minutes and then add one cup of plain yogurt, stirring it in thoroughly. Keep the temperature at 105 degrees for ten minutes, then turn off the heat.

» I'll add the yogurt once the milk reaches 105 degrees. You could also use commercial thermophilic starter culture for this step.

>> Here I am adding the dissolved rennet by pouring it slowly through a slotted spoon for better distribution.

Once the temperature has dropped to 95 degrees, add the rennet by sprinkling it over the milk and mixing using a gentle up-down and back-and-forth motion. Remove the pot and cover it with the lid. Allow the mixture to set for about an hour and then check for the development of the curd. Check the curd by inserting a clean and sterile blunt object (such as a glass candy thermometer). If it can be withdrawn cleanly without anything sticking to it, and the hole it makes doesn't immediately fill with liquid, the curd is ready and you have what is called a clean break. If the curd isn't ready, allow the pot to set while covered for another fifteen minutes and check again.

Now that you have a clean break, you need to cut the curd. The purpose of cutting the curd is to allow for uniform drainage of the milk liquid (known as whey) from the curd. (Yes, this is the famous "curds and whey"—a primitive predecessor to cottage cheese—likely eaten by Miss Muffet in the nursery rhyme.)

Your goal in cutting the curd is to cut the curd into uniform-sized curds for even drainage of whey. In general, the smaller you cut the curds initially, the harder the style of cheese you are making; though there are practical limits. In this case, you are cutting the curd into one-inch cubes. Do this by using your curd knife to first cut a grid at right-angles the entire depth of the curd so you end up with a one-inch checkerboard pattern. Then, make horizontal cuts by positioning your curd knife at a 45 degree angle and cutting along one row of parallel lines in your grid. Though there are all sorts of other ways to do this and special gear you can buy, it is really that simple.

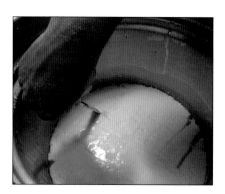

Once your curd is cut, cover the pot again and allow it to sit for another fifteen minutes so some whey can gather at the bottom of the pot. Then, put your pot back into the double boiler and slowly, over a period of 30 minutes or so, raise the temperature of the

◀ The horizontal cuts are being made by tracing the grid with the knife held at a 45 degree angle.

» The curds will release whey and shrink. The metal device is the temperature probe.

curds to 110 degrees. As the curds are heating, gently—very gently so you don't break them—use your slotted spoon to stir the curds in such a way as to exchange those on the top with those on the bottom in order to promote even heating. Once the curds have reached 110 degrees, keep at that temperature for thirty minutes while gently mixing every five minutes or so. You will notice the curds getting smaller and the amount of whey increasing. While this process is ongoing, prepare a large piece of cheesecloth by boiling.

Line your colander with a double-layer of cheesecloth, and gently pour the curds and whey into the colander. You can save the whey for baking later, add it to your compost pile or let it go down the sink. (If the whey is greenish, do not be alarmed—this is normal!) Let the curds drain in the colander for an hour or so, then put the curds into a bowl and salt to taste, turning the curds evenly for uniform distribution. I prefer sea salt for this, but you can also use cheese salt or canning salt. Do *not* use regular table salt (iodized or not) because it will make your cheese taste bitter.

⊗ The cheese is being mixed with flaked cheese salt.

Prepare some more coarse cheesecloth by boiling, and then use a double layer to line your clean cheese mold. Add the curds to the mold, fold the cheesecloth over top of the curds, and put the top of your mold on top of the cheesecloth. Put your mold in a shallow pan (a disposable pie plate would be ideal) to catch whey that is expelled. Add two pounds of weight on top of the mold, and place the whole works in the refrigerator.

» I used a 2.5 pound weight on the cheese press, and it worked fine.

Practical Cheese Making Techniques

❯❯ The completed cheese before wrapping it in plastic and storing in the refrigerator.

Once the cheese and press have been allowed to work in the refrigerator for four or five hours, turn the cheese out of the mold, unwrap it, and place in a closed container in the refrigerator. Use within a week.

Hard Minimally Aged Cheese Using Mesophilic Starter: New Ipswich Jack

Mix one teaspoon of 30% calcium chloride solution into a quarter cup of water, and mix with two gallons of milk in a double-boiler. Bring the temperature of the milk up to 85-90 degrees, and add either ½ tsp of powdered mesophilic starter or one cup of fresh cultured buttermilk, mixing thoroughly. Cover the mixture and allow it to ripen for 30–40 minutes while maintaining the temperature between 85 and 90 degrees.

While the mixture is ripening, prepare your rennet solution by mixing ½ tablet of rennet with ¼ cup of cool non-chlorinated water. You'll know the mixture has ripened by the fact 30–40 minutes have passed and it smells like buttermilk or yogurt. Once the mixture has ripened, add the rennet solution by dripping it around the milk and mixing it gently but thoroughly using up-down and back-and-forth motions. Continue to maintain a temperature of 85 to 90 degrees while allowing the mixture to sit covered for an hour. At this point, the curds should give you a clean break.

Use your curd cutting knife to cut the cubes into 1/4-inch cubes. Continue holding the temperature at 85 to 90 degrees for another 40 minutes while gently stirring the curds every five minutes or so. Keep the curds covered while not stirring or checking the temperature. You'll notice the curds shrinking and the volume of whey increasing.

Slowly increase the temperature to 100 degrees over a 30-minute period while stirring every five minutes or so. This amounts to about two degrees every five minutes. Hold the temperature at 100 degrees for another 30 minutes while stirring every five minutes.

Now, very gently so as not to damage or lose curds, pour off as much of the whey as you can. This may be easier to do with a helper holding back the curds using the slotted spoon while someone else tips the pot over the sink.

Put the pot back into the double boiler and continue to stir for another 30 minutes while maintaining the temperature at 100 degrees. Meanwhile, prepare a double-layer of course cheesecloth by boiling first, and use it to line a colander. Pour the curds into the cheesecloth-lined colander. Add two tablespoons of cheese salt and mix the curds gently.

Line your cheese mold with cheesecloth, and then pack the mold closely with the curds. Fold your cheesecloth over top of the curds, install the top of your mold, and put your mold in a shallow pan to catch the whey that will be expelled.

Put a ten-pound weight on top of the mold to press the cheese for fifteen minutes. Then, remove the cheese from the mold, take it out of the cheesecloth, flip it over within the cheesecloth, and put it back in the mold.

This time, press the cheese for 30 minutes with a thirty-pound weight. (I recommend stacking three 10-pound dumbbell weights as these are easier to handle.) Then, take the cheese out of the press, take it out of the cheesecloth, flip it again within the cheesecloth, re-cover it, and put it back in the mold. Press it this time with 40 pounds for twelve hours.

After this, take it out of the mold and cheesecloth, and lay it on a bamboo sushi rolling mat. Flip it on the mat once a day so that it dries evenly. After three to five days, it should be dry to the touch. Once it is dry to the touch, it is ready for aging.

This cheese should be aged at temperatures of from 50 to 60 degrees for anywhere from one to three months. Maintaining such temperatures is a tall order in most homes, but any temperature range from 45 to 68 will do. Luckily (at least in this respect) my house is old and drafty so I can age cheese in a kitchen cabinet anytime from November to April without need of maintaining a special environment.

If, however, you happen to either live in a warmer climate or have a more energy-efficient home, you will likely need to create a cheese-cave from a dorm refrigerator as described earlier in this chapter.

Larger cheeses will form a natural rind that will protect them from invasion; smaller cheeses (like the size that we have made in this example) will need to be protected by either wax or plastic.

If using plastic, first wash the cheese using vinegar on a clean paper towel to reduce bacterial counts, then seal it in plastic using a vacuum sealer.

If you are using cheese wax, melt it by putting a small stainless steel bowl in a pot of boiling water and adding wax to the bowl. (This bowl will be almost impossible to clean after, so you might want to get a cheap bowl at a department store for this purpose.) After you have washed the cheese with vinegar, use a natural bristle brush to dip in the melted wax and then paint it onto the cheese. Once the wax has hardened on one side of the cheese, turn the cheese over and coat the other side. Check the cheese over thoroughly to make sure you haven't missed any spots and that the cheese is coated uniformly, and then set the cheese aside to age.

After this cheese has aged for a month, it is safe for people who are lactose intolerant; after it has aged for two months, it is safe even if made from raw milk.

Monadnock Cheddar

This cheese starts off identically to the New Ipswich Jack cheese, and the primary variance starts with the cheddaring process. Mix one teaspoon of 30% calcium chloride solution into a quarter cup of water, and mix this with two gallons of milk in a double-boiler. Bring the temperature of the milk up to 85-90 degrees, and add either ½ tsp of powdered mesophilic starter or one cup of fresh cultured buttermilk, mixing thoroughly. Cover the mixture and allow it to ripen for 45-50 minutes while maintaining the temperature between 85 and 90 degrees.

While the mixture is ripening, prepare your rennet solution by mixing ½ tablet of rennet with ¼ cup of cool non-chlorinated water. Once the mixture has ripened, add the rennet solution by dripping it around the milk and mixing it gently but thoroughly using up-down and back-and-forth motions. Continue to maintain a temperature of 85 to 90 degrees while allowing the mixture to sit covered for 45 minutes. At this point, the curds should give you a clean break. If not, allow to sit for another 15 minutes and test again.

⊗ The cut curds are starting to release whey.

Use your curd cutting knife to cut the curd into 1/2-inch cubes. Slowly increase the temperature to 97 to 100 degrees over the next 40 minutes while gently stirring the curds every five minutes or so. Keep the curds covered while not stirring or checking the temperature. You'll notice the curds shrinking and the volume of whey increasing. Hold this temperature for another 30 minutes while stirring periodically to prevent matting or clumping. During the last five minutes, don't stir so the curds can settle on the bottom.

Line a colander with boiled cheesecloth and put the colander over a large pot to collect the whey. Pour the contents of your double-boiler into the colander and put a lid on top to retain heat, and allow to sit for one hour. (Make sure that the level of the whey in the pot isn't high enough to actually touch the curds. Any excess whey can be drained or put in your compost pile.)

Here is the the distinctive process that makes cheddar cheese and it is called "cheddaring." At the end of an hour you'll notice that the curds have amalgamated into a solid mass. Cut the mass of curd into slabs about ¼ inch thick. Stack the curds like dominoes and cover with the cheesecloth. Every fifteen minutes, rearrange the stack so the slabs that were outside are now inside, and those that were on the top are now at the bottom. After four or five rounds of this, the texture should resemble very firm tofu or turkey breast.

⊗ The slabs are cut and being stacked.

⊗ The cheese slabs are stacked. You can see the liquid draining from them.

⊗ Cheese in the process of milling.

After the cheese slices have reached the desired texture, they need to be milled. That just means you need to cut these slices up into chunks about the size of a pea. Put the milled curd in a bowl and sprinkle with 1 ½ teaspoons of flaked salt while gently rolling the curds around for uniform distribution.

Line your cheese mold with cheesecloth, and then pack the mold closely with the curds. Fold your cheesecloth over top of the curds, install the top of your mold (also called the follower), and put your mold in a shallow pan to catch the whey that will be expelled.

Put a ten-pound weight on top of the mold to press the cheese for fifteen minutes. Then, remove the cheese from the mold, take it out of the cheesecloth, flip it over in the cheesecloth, and put it back in the mold. Press with the ten pounds of weight for another 45 minutes. The reason we use a light weight at first is to prevent expelling fat with the whey.

Then, flip over the cheese in the cloth, and increase the weight to 40 pounds for 24 hours. The next day, flip the cheese again, and continue to press with 40 pounds for another 24 hours. So in total, the cheese has been pressed for two days and an hour.

Now, remove the cheese from the cloth, wipe it down with either vinegar or a brine solution using a clean cloth, and place it in a protected room-temperature place on a bamboo mat for a day or two until a rind starts forming. (You can make

⊗ The pressed cheese wheel on a bamboo mat.

⊗ A wire scaffold will hold the cheesecloth away from the cheese and keep flies away.

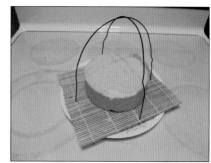

brine solution by mixing as much salt into water as the water will dissolve.) What I do to keep insects away is put a wire scaffold over the cheese and drape some cheesecloth over it. Once the rind has started to form, either coat the cheese with cheese wax or seal it in a plastic bag using a vacuum sealer.

Mature the cheese from one to three months. The ideal temperature for aging cheddar is 50 degrees. During the winter, this is the temperature of my porch so I'm lucky. You can age it at temperatures ranging from 45 to 60 and it will come out fine.

Tip for Maintaining Temperatures

Reading books about cheese making, you'd think everyone owns a precisely controlled stove that allows maintaining temperatures within a single degree for hours on end. In the real world, maintaining temperatures is somewhat difficult. Using a double-boiler keeps the milk from scorching. Unfortunately, when you raise the temperature of the milk to, say, 90 degrees using a double boiler, if the inner container is left in the outer container the temperature will continue to rise well beyond that of the culture you are using even if you turn off the heat.

Once the milk has reached the higher end of the temperature range, simply remove the inner container from the outer container and set it on an unused burner on the stove. Check the temperature once in a while and if it seems to be going too low, just set it back in the outer container for a few minutes, and take it back out once the temperature is in the proper range. Because the amount of milk and water involved has a substantial thermal mass, usually this need not be done more than once for a particular waiting period.

Experiment and Keep a Log

A lot of times people want to make cheeses like those that they buy at the store. If you want to do that, there are a host of sites on the Internet that give specific recipes. But what I recommend instead, is that you experiment and keep a log. I have covered all the fundamental principles you need to know in order to make your own unique cheeses. Fresh cheeses have to be refrigerated to be safe and should be used in less than a week. Cheeses made from raw milk have to be aged for at least two months to be safe. Hard cheeses need to be pressed with increasing amounts of weight. But now, from just the four cheeses I have given in this chapter, you can think about the variations.

The Queso Blanco recipe was a direct acidification cheese made with vinegar. What would happen if, instead of adding vinegar, you added a mesophilic starter and held it at 88 degrees for an hour before pouring into the cheesecloth? It would certainly taste different!

The soft Farmer's Cheese described earlier used a yogurt (thermophilic) starter culture. What if you used the same technique, but instead used a buttermilk (mesophilic) starter and varied the temperature accordingly?

The Jack cheese recipe is pretty interesting. Don't you wonder what would happen if you used a thermophilic starter and some lipase instead of a mesophilic starter? How would it come out? What would it taste like? What would happen if you added a pint of heavy cream and a tablespoon of wine vinegar to one of the recipes?

So rather than copying other recipes, what I am encouraging you to do is follow the general principles I have described here to make your own and keep notes. I think you will be very pleasantly surprised at how easy it is to make astonishingly good cheese that is uniquely your own and can't be bought anywhere at any price. This is ultimately what will make cheese-making a worthwhile thing for a mini-farmer.

PART VI

Bread for Every Occasion

15

Artisan Breads on the Stone

Though the wisdom of consuming grains in any form has been credibly challenged by some biochemists,[29] [30] bread remains a staple food for millions if not billions, and is known as the staff of life. Bread in some form is present in nearly every culture and the shapes it takes are myriad.

Some of the most cherished breads in the Western world are what are now known as artisan breads, and you can buy a one-pound boule in the supermarket for around six dollars. Tender on the inside, crunchy on the outside, and indescribably delicious, it's a shame that such bread is priced out of reach of the average person for anything but special occasions. But it doesn't have to be.

29 Cordain, L. (2010), *The Paleo Diet, Revised Edition*
30 Wolf, R. (2010), *The Paleo Solution: The Original Human Diet*

One benefit of a self-sufficient approach to living is that it gives you access to a lifestyle that would be unaffordable otherwise. When you do things yourself, you get to have foods of the highest quality for less money than inferior items. You can make artisan bread yourself for about fifty cents a loaf.

The Basic Chemistry of Bread

Though other grains are often used in bread, wheat is the ideal because of its protein structure. Wheat flour contains starch and five protein groups—albumin, globulin, proteoses, gliadin, and glutenin. Of these, only the latter two aren't soluble in water. When flour and water are mixed, the albumin, globulin, and proteoses are dissolved. This gets them out of the way and the gliadin and glutenin combine to form gluten.

Gluten is what gives bread dough its elasticity. If you've ever seen pizza crust made, you've seen how far it can stretch. In order for gluten to form chains of that sort, the dough can either be relatively dry and kneaded, or it can be relatively wet and allowed to sit for a few hours.

Artisan bread has only four basic ingredients: flour, water, salt, and yeast. That's all. Nothing more is needed. Bread made in bread machines needs to rise rapidly, so sugar is included so the yeast will have immediate access to food. Because the dough for artisan breads is allowed to sit, during which time a certain amount of autolysis occurs, some of the starch in the flour is naturally converted to sugar.

Salt is used in bread for two purposes. The first is to limit the activity of the yeast so you don't wind up with huge air gaps in your bread. The second is to strengthen the gluten. The yeast used for bread is the same species as that used for wine and beer, but the specific variety has been selected for baking purposes. The yeast eats sugars and makes alcohol and carbon dioxide. The carbon dioxide makes the bubbles in the bread, and the alcohol evaporates during baking.

A potential fifth ingredient can also be included, and that is lactobacillus— lactic acid bacteria. Lactobacillus can live symbiotically with bread yeast. When it does, it turns the alcohol byproduct of yeast into lactic acid, which gives sourdough bread its flavor. The lactic acid helps to preserve the bread and gives it a shelf life that is nearly as long as that of commercial breads containing preservatives. Up until

❯ Artisan bread requires only these four basic ingredients.

the 1800s, practically all leavened bread was sourdough because yeast and lactobacillus as separate organisms were unknown. Once the difference was discovered, yeast was cultured by itself for the purposes of leavening. So, interestingly, by separating the symbiotic yeast/lactobacillus culture for convenience, preservatives in bread became necessary.

The Five-Minute Method

Artisan breads have traditionally been time-consuming to make, but the combination of two innovations allows you to make no-knead bread in as little as five minutes a day. The first innovation was introduced in 1994 by Jim Lahey of the Sullivan Street Bakery in New York City. This was the incorporation of a substantially larger proportion of water into the dough and allowing longer sitting times. This allows the gluten chains to link without kneading. The second innovation was introduced in 2007 by Jeff Hertzberg and Zoë François, and consists of the simple fact that dough made in this fashion can be refrigerated. When the two innovations are combined, you can make delicious artisan bread in mere minutes. The following is my explanation and adaptation of the method.

The core idea of the method is that if you make a very wet dough and set the dough aside in the refrigerator, the gluten chains will interlink on their own over time, thus obviating the need for kneading the bread to obtain a good consistency. The dough can be kept in a covered container in the refrigerator for up to two weeks, and all you have to do is take it out, cut off a portion of it, let that portion rise, and then pop it in the oven. Over time, as you save portions of the dough from previous batches for your new batches in the same bowl, your bread will develop its own sourdough character without need for maintaining separate sourdough cultures.

Equipment

To use this method you need three pieces of equipment you might not otherwise have. The first is a baking stone. A baking stone is a thin round stone on a wire rack. Many stores call it a pizza stone. Because of its nature, it absorbs more moisture from the bread, thus making a crisper crust while also preventing burning. You can buy either glazed or unglazed baking stones, but for artisan bread you want an unglazed stone. An unglazed baking stone should be seasoned by coating it with oil periodically and should only be rinsed in lukewarm water and wiped off. If you were to use soap, it would harm the seasoning and the soap would be

absorbed into the pores. I use an unglazed stone I picked up at a department store for about $15.

Second, you need a broiler pan. Put the broiler pan in the oven on the rack underneath the one holding the baking stone and bread. Put some water in the broiler pan so that steam gets trapped in the oven with the bread and the crust won't dry out too much.

Finally, you need a pizza peel. A pizza peel is an oversized spatula that you use to put the bread onto the hot stone, and to remove the bread from the hot stone. For a long time I just used a big spatula, but a peel works better. A simple wooden peel for $10 works fine.

Dough

The dough used in this method has more water than typical yeast bread. Yeast bread most often includes water and flour in a 1:3 ratio, whereas the dough used in this method has a water-to-flour ratio of about 1:2. The reason is because the higher water content makes the gluten more mobile, and this allows it to form long elastic strands without kneading. The dough also uses less yeast, because the longer sitting times give the yeast more time to multiply. The dough is made using what the authors term the 6-3-3-13 rule.[31] Enough dough for eight loaves of bread can be made with 6 cups of water, 3 tablespoons of yeast, 3 tablespoons of salt and 13 cups of flour.

It is important to measure flour properly. You use a dry measuring cup, spoon the flour from the bag into the measuring cup, and then level it with the back edge of a knife. The difference in the

» Spoon flour from the bag into the dry measuring cup and then level it with the back edge of a knife.

---

31 Hertzberg, J. & Francois, Z. (2008-2009), "Five Minutes a Day for Fresh Baked Bread," *Mother Earth News*, December 2008—January 2009

» The dough is uniformly moistened.

weight of flour in a cup can be as much as 40% when using different techniques, so consistency of measuring technique is important.

The technique for mixing the dough is to put the water in a large bowl, mix in the yeast and salt, then incorporate the flour until it is uniformly moist. You can use a dough hook on a mixer, or your hands. Once the dough is thoroughly moistened, leave it covered with a kitchen towel to rise for two or three hours. Then, cover the dough with plastic wrap and put it in the refrigerator. The dough can now be used anytime from three hours to two weeks after making.

When you run out of dough and need to make more, don't wash out the old dough, just incorporate whatever is left over, whether a tiny bit or a good handful, into the new dough. This way, over time, your bread will accumulate a unique sourdough flavor that is missing from the first few loaves.

⊗ The dough before being placed in the refrigerator.

⊗ The proper amount of dough for a loaf of bread.

Making the Bread

It literally couldn't get any simpler. Flour your pizza peel so the dough won't stick to it, take out your container of dough and use a serrated knife to cut off a hunk large enough to fill both hands. If your hands are smallish, use enough to fill your hands twice.

Flour the dough so it doesn't stick and shape it into a ball. (Bread in this shape is called a "boule.")

» Shaping the boule by stretching the dough from the top around the sides to the bottom.

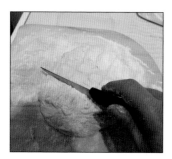

⦉ The shallow slashes allow the bread to expand without cracking.

⦊ Sliding the boule onto the pre-heated baking stone.

Then, put it on your pizza peel to rise and put a few shallow slashes in the top to allow for expansion during baking. Set your timer for twenty minutes.

At the end of twenty minutes, put your baking stone in the oven and pre-heat the oven to 450 degrees and put the empty broiler pan on the rack under the stone. Set your timer for another twenty minutes. All together, this will have allowed forty minutes for the dough to rise.

When the time is up, pour a cup of water into the broiler pan and slide the bread dough from the pizza peel onto the baking stone. You may need to use a bit of a jerking motion similar to what you see the cooks doing in a pizza restaurant.

Close the oven and set your timer for 30 minutes. At the end of 30 minutes, remove your bread from your stone and put it on a wire rack to cool. I've found you can bake it for up to three minutes longer for a darker crust without danger of drying it out.

Though there is some wait time with this technique during which you can do other things, your actual hands-on involvement is only five minutes. This is even faster than using a bread machine!

Variations

Frozen Dough

Instead of keeping the dough in the refrigerator you can freeze it in one-pound chunks. Allow the chunks to defrost overnight in the refrigerator before using them. This is handy if you don't eat much bread but want to have it available with minimal fuss.

⦿ Cooling on a wire rack preserves the crunchiness of the bottom crust.

Different Flour

All-purpose flour is what the recipe usually uses, but you can substitute bread flour or a portion (up to half the total flour) with whole wheat flour for variations in taste and texture. The recipe is pretty forgiving.

Sourdough

Using this method, over time your bread will develop its own sourdough culture. But you can use the instructions in the next chapter for making a sourdough culture, and substitute three cups of sourdough culture for one of the tablespoons of yeast.

Different Shapes and Added Ingredients

Dough is dough. By incorporating other ingredients or shaping it differently, you can use it to make cinnamon buns, donuts, dinner rolls, pizza crust, and more.

Pizza crust is just a matter of shape. Cut off a one-pound piece of dough, coat it with flour so it doesn't stick, and roll it out. Allow to sit twenty minutes. You can keep it on a floured pizza peel as you add the other ingredients, and then slide your pizza onto the baking stone in a preheated 450 degree oven. Cook for 15–20 minutes, until the cheese is golden brown.

Cinnamon buns require the addition of some sugar, fat, and cinnamon. Take a one-pound piece of dough, use your hands to incorporate two tablespoons of cooking oil, four tablespoons of sugar, and ½ teaspoon of cinnamon. Roll out flat, cut into ribbons, and then roll up the ribbons into close spirals. Sprinkle with sugar and cinnamon, allow to rest for twenty minutes, and then cook for twenty minutes in a preheated 450-degree oven.

Donuts require the addition of four tablespoons of sugar per pound of dough and any other spices you prefer. Roll the dough into one-inch balls or shape them like regular donuts if you are more ambitious. Allow to rise for twenty minutes. Deep fry at 350 degrees for fifteen minutes. Take out, drain, and serve warm.

Dinner rolls with this dough aren't as light and fine-grained as with a standard sweet roll recipe, but they are still quite good. Roll the dough into two-inch balls, place on a cookie sheet, pat down a bit, let rest for forty minutes, then bake for twenty minutes in a preheated 400-degree oven.

16

Bread on-the-go with Your Bread Machine

If you like bread, about the coolest device in the world is a bread machine. I got my first one at a pawn shop for $25 complete with a manual, and some of my daughter's earliest memories include eating peanut butter on freshly-sliced hot bread from the bread machine.

Naturally, the bread from a bread machine contains far fewer questionable ingredients than most bread from the store. It certainly doesn't contain the wide array of emulsifiers, preservatives, dough conditioners, and similar chemicals necessary for making and distributing bread on an industrial scale. In addition, you can control the quality of the ingredients by using organic butter, whole wheat flour, and so forth. The results taste dramatically better than store bread.

Many people buy bread machines with the best of intentions, but after a couple of uses they wind up forgotten in the

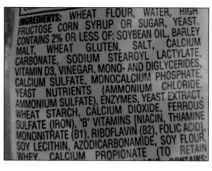

> The most powerful argument for making your own bread is the ingredient label on the bread at the store.

back of a closet somewhere. There are a couple of reasons for this that are easily remedied.

The first reason is that even though a bread machine does away with a lot of the work involved in making bread, it can take twenty minutes to gather the ingredients and another five or ten to put them in. I realize that doesn't sound like much, but the fact is that time is in short supply for most families and a thirty-minute project is about twenty five minutes too long.

The second reason is that bread machine bread is hard to cut into uniform slices for sandwiches. Inexpensive fixtures are manufactured that make slicing your bread a breeze, but such fixtures are never included with bread machines and they are seldom if ever displayed in stores where they are easy to find. The lack of ability to uniformly slice bread from the machine limits its utility.

As a result, bread machines are too often reserved for special occasions, and store-bought bread is used day-to-day for sandwiches and the like.

In this chapter I am going to describe a very simple method that will allow you to make bread anytime in just five minutes and point you in the direction of guides to help you slice your bread. In addition, I am going to explain how to make and use sourdough starter along with the ingredients and principles used to formulate bread machine recipes. This way, you will be well-equipped to make your own recipes even if you bought your bread machine without a manual at a yard sale.

Bread Machines

Since their invention in 1986, there has been a profusion of makes and models of bread machines with a wide array of features. Over time, their relative cost has dropped as features have increased. My current bread machine even has cycles for making jam, among other things. You can spend a lot of money on a bread machine—as much as $300. But unless you are using it for specialized breads, you will do fine with a $100 model.

In general, a bread machine has timed cycles for preheating ingredients to optimal temperatures, mixing the dough, and for rising, punch-down, and baking.

» This inexpensive bread machine has more features than you'll ever need.

How these are timed depends on the type of bread you are making and how dark you'd like the crust. Most machines have provision for easily selecting the proper settings for the more common types of bread.

Bread machines have short cords. This is allegedly (according to one user manual I read) to keep people from tripping on them. In practice, it means you have to locate them near an electrical outlet. They contain a non-stick pan, and in the center of the pan is a paddle for mixing the dough. Sometimes the paddle gets enmeshed in the bread so that it can't be easily removed without losing a small chunk of bread, but spraying it with a bit of non-stick cooking spray before putting in the ingredients helps. Ingredients are added to the bread pan with liquids first, then mixed dry ingredients, yeast, and oil (or butter) in that order.

The one feature in a bread machine I believe is most useful is the ability to start the bread-making process at a future time. Even once all the ingredients are in the pan, it can easily take over three hours for the bread to be finished. If you get home at 6PM, take care of a few things and then start the bread at 6:30PM, it will be 10PM before the bread is ready. That just doesn't work. You need the ability to dump the ingredients in the pan and push a few buttons before you head off to work and arrive home to the smell of bread already baking. Most bread makers have this feature, but check before you buy one.

Ingredients

Artisan bread uses just four ingredients: flour, water, salt, and yeast. Bread made in bread machines also includes oil, sugar, powdered milk, and sometimes eggs. To some extent the purposes of these ingredients is well-known, but I'd like to expand on them a little.

Flour

Wheat flour is available in self-rising, all-purpose, bread, and whole wheat varieties. Avoid self-rising flours for making yeast breads as the leavening they already contain would most certainly ruin the bread. Though bread flour will make a finer-textured bread and it has a bit more gluten protein, it isn't necessary for making bread, although some recipes will specify it. All-purpose flour is what is most commonly used for … well … all the things for which flour is used, including making bread. Whole wheat flour is flour made from the whole wheat kernel.

Flour is the heart of bread because it is what forms the dough. And the gluten in wheat flour is what makes leavened bread so delightful. Other grains contain gluten to various degrees—barley and rye particularly. But in any bread (except for gluten-free breads) wheat flour will be the core ingredient with any other flour serving only as an adjunct.

Water

You want to use good water for making your bread. If your water is bad, filter it with reverse osmosis or use bottled water.

Salt

Salt is used in bread for two purposes. The first is to limit the activity of the yeast so you don't wind up with huge air gaps in your bread. The second is to strengthen the gluten. Though many recipes and authors state that non-iodized salt is best, my results have been the same using sea salt, Celtic sea salt, iodized salt, and non-iodized salt. The important thing is merely that the primary ingredient in your salt should be just plain salt: sodium chloride.

Yeast

The yeast used for bread is the same species as that used for wine and beer, but the specific variety has been selected for baking purposes. The yeast eats sugars and makes alcohol and carbon dioxide. The carbon dioxide makes the bubbles in the bread, and the alcohol evaporates during baking. You can use bread machine yeast, rapid rise yeast, or ordinary baking yeast. All work equally well, and the only difference is that you need less of the rapid rise or bread machine yeast than you would need of ordinary baking yeast. The following table lets you interchange them in recipes.

Ordinary Baking Yeast	Rapid Rise or Bread Machine Yeast
¾ tsp	½ tsp
1 tsp	¾ tsp
1½ tsp	1 tsp
2¼ tsp	1½ tsp
1 tbsp	2 tsp

Sugar

Because dough used in bread machines isn't as wet as that used for artisan breads and it doesn't have enough time to autolyze, the yeast needs sugar in order to leaven the bread. Any sugar not eaten by the yeast contributes to an aesthetically pleasing golden-brown crust. You can use honey or molasses as a substitute for sugar, but don't try to use artificial sweeteners because yeast can't process them.

Fats

Fats help make bread and crust more tender and help the bread stay fresh longer. Usually, either a high-quality vegetable oil or butter is used. If you use butter straight from the refrigerator, cut the requisite quantity into small chunks so that the bread machine can blend it into the dough more easily.

Eggs

Eggs are used to give breads a soft, velvety texture. When eggs are specified in recipes, large eggs should be used. Also, before an egg is used, allow it to sit at room temperature for an hour or so. Eggs should never be used in conjunction with a delay timer as allowing them to sit several hours in the bread pan will cause salmonella or other dangerous bacteria to develop. Only use them in recipes you are starting immediately.

Sourdough

Sourdough is a type of leavening consisting of symbiotically-paired lactobacillus and yeast. Every sourdough is a little bit different, and some cultures have been passed down across generations since the 1700s or before. Sourdough starter is used in place of yeast, but it can also be used in tandem with yeast.

Because wild yeasts and lactobacillus are ubiquitous in our environment, it is entirely feasible to make your own sourdough completely from scratch, and I'll cover that later in this chapter. You can also buy and propagate commercially-made sourdough cultures.

When you are using sourdough culture in place of yeast in a recipe that will be delayed in starting, mix the sourdough with the water at the bottom of the bread pan rather than putting the culture on top of the flour as you would do with dried yeast.

Culturing and Maintaining Sourdough

As I've mentioned, you can readily make your own sourdough. The procedure is easy enough, and it makes sense if you plan to make bread at least once a week. If you don't plan to make bread that often, then instead of using real sourdough you can get something that tastes similar by adding ¼ tsp of citric acid (which you may already have for making wine!) to the ingredients of the loaf.

To make your own sourdough culture, get a four-quart wide-mouthed glass container. Add two cups of water, 3½ cups of flour, 2¼ teaspoons of active dry yeast, 1 tbsp granulated sugar, and stir until smooth. Cover loosely (not airtight) with plastic wrap, set in a warm place (70–80 degrees) for five days stirring three times a day. Over time, the starter will grow, shrink, become bubbly, get thinner and even develop a yellow liquid layer on top. Just keep stirring it three times daily. At the end of five days, it should be obvious that you have an actively fermenting starter; cover with plastic and put in the fridge. It's now ready to use.

This is a little bit hit-or-miss. You will usually end up with a sourdough starter, but how good it tastes is anyone's guess until you try it. You may or may not get lucky. If you don't, trying again is inexpensive. Every week, feed it a teaspoon of granulated sugar. Whenever you use a cup of sourdough starter, replace with a mixture of ⅓ cup water, ½ cup flour and 1 tsp sugar.

If you plan to use sourdough a lot, you may benefit from buying a commercial starter culture. You can get good commercial cultures from King Arthur Flour[32]

---

32 King Arthur Flour, 1-800-827-6836, www.kingarthurflour.com

and Sourdoughs International.[33] Another useful source of sourdough culture is Friends of Carl.[34] Friends of Carl is a group of people who have continued Carl Griffith's tradition of providing sourdough culture for free to anyone who asks. Follow the instructions on their website to get a sourdough culture dating back to 1847.

If you buy a commercial starter culture (or obtain Carl's), follow the directions with the culture for starting and maintenance rather than my generic directions. Each culture is acclimated a particular way, and my way may not work best for certain cultures.

Principles and Patterns in Bread Machine Recipes

When I first got a bread machine, the recipes seemed rather mysterious. The instructions emphasized the care that must be exercised in measuring, as though a single errant molecule would wreck the bread. But if you look at something long enough, patterns and rules emerge, and bread machine recipes are no different.

Bread machine recipes maintain a ratio of 1:3 between water and flour. This is the case even if a recipe uses half whole-wheat flour and half regular or bread flour. You can vary this slightly for a wetter dough (e.g. increase the water by ¹⁄₁₆ or decrease the flour by ⅛), but you shouldn't make a drier dough. You'll want to increase the volume of water by ¹⁄₁₆ if you use all whole grain.

Salt is used at a ratio varying from 0.4–0.5 teaspoons of salt per cup of flour. Except in recipes such as French bread that use no oil, the amount of oil used is identical to the amount of sugar used, and the amount of sugar varies from 2–3 teaspoons per cup of flour. When powdered milk is used, it is used at a rate of one to one-and-a-half teaspoons per cup of flour. The amount of bread machine or rapid rise yeast equals the amount of salt used. If sourdough starter is used, one cup is used to replace 1½ teaspoons of rapid-rise yeast, and ⅓ cup is subtracted from the amount of other non-oil liquids used. Liquid milk, if used, is substituted one for one with water for up to half the liquid.

That's all. The only other constraint is that bread machines are rated by the weight of the loaves they can handle. Their baking cycles are calculated on the

---

33 Sourdoughs International, 1-208-382-4828, www.sourdo.com
34 Friends of Carl, Oregon Trail Sourdough P.O. Box 321 Jefferson, MD 21755 USA, http://carlsfriends.net

basis of loaves of a certain weight, and the recipe has to be compliant with those sizes. The basic form of bread machine recipes can be summed up in the following table.

Ingredients	Proportions	One Pound	1.5 Pound	2 Pound
Water	1 cup	⅞ cup	1 cups	1⅓ cups
Flour	3 cups	2 ¾ cups	3 cups	4 cups
Salt	1.2 tsp to 1.5 tsp	1.5 tsp	2 ¼ tsp	1 Tbsp
Rapid-rise Yeast	1.2 tsp to 1.5 tsp	1.5 tsp	2 ¼ tsp	1 Tbsp
Sugar	2 Tbsp—3 Tbsp	3 Tbsp	¼ cup + 1 tsp	¼ cup + 2 Tbsp
Oil	2 Tbsp—3 Tbsp	3 Tbsp	¼ cup + 1 tsp	¼ cup + 2 Tbsp
Powdered Milk	1 Tbsp—1.5 Tbsp	1 Tbsp	4.5 tsp	2 Tbsp

Bread in a Bag

In *Maximizing Your Mini Farm,* I described a method for setting aside frozen meals in such a way as to avoid the need for eating out. Such a method works because preparation time and cooking time are about the same without regard to quantity. The difference in time for making four steaks instead of two is minuscule, so just make four so you can put the two you aren't eating today in the freezer. If you do the same with vegetables, in no time flat you'll have a stock of ready-made items that you can mix and match for a meal by just popping them in the microwave. This is really easy, saves time, and most importantly saves money, because people spend as much as half their food budget eating out due to a perceived lack of time to make meals.

The same is true of making bread with a bread machine. Most of your time is spent in gathering ingredients, gathering the proper measuring tools, and looking back and forth at the recipe. It doesn't seem like much, but it takes the average person about twenty minutes—and this might be twenty minutes they don't have.

So instead of making up ingredients for a single loaf of bread, make up all the dry ingredients for four or eight loaves of bread and seal them in individual bags. This will take about forty minutes instead of twenty. But the per-loaf cost in terms of time would drop from twenty minutes a loaf to five minutes a loaf.

All you do is set up four to eight bags, and then add in all the dry ingredients. First put ingredient X in all eight bags, then put ingredient Y in all eight bags, etc.

⊗ Sliced bread is better for sandwiches. You can adjust the thickness by skipping slots.

until done. Put all of the dry ingredients but yeast in all the bags, seal them, and shake them up. In this way, you get rid of the need for pre-mixing the dry ingredients each time you make a loaf of bread.

In addition to your homemade bread mix, you usually only need three ingredients: water, butter, and yeast. That's all. When you are ready to make bread, put the measured amount of water in the bread pan, dump in your mix, put the butter and yeast on top, close the cover, and turn it on. Come back later and enjoy your bread!

As Good as Sliced Bread

All you need to slice bread effectively, even into very thin slices, is a bread slicing guide and a slicing knife. A bread slicing knife is designed with the specific objective of slicing bread, and it does so better than any other sort of knife. They aren't expensive, so if you eat bread, it is worth buying one.

There are a few models of bread slicing guides available. I use a slicing guide called "BreadPal" but you could buy other versions or even make your own. The picture shows how straightforward it is.

The key to slicing bread is to let it cool first. Once it is cool, it slices straighter and with less tearing. Take your cooled loaf, put it in the slicing guide, and use your bread slicing knife to slice the bread. Now you have bread that is easily toasted, used for sandwiches, and all the other uses for which store-bought bread is often preferred.

Hannah's Half Whole Wheat

Earlier in the chapter I mentioned my daughter Hannah enjoying bread from the bread machine. Here is the recipe for a 1.5-pound loaf of her favorite bread:

Ingredients

1 cup + 1 Tbsp	Warm water
1 Tbsp	Honey
1 ½ Tbsp	Packed brown sugar
1 ½ cups	Bread flour
1 ½ cups	Whole-wheat flour
1 ½ tsp	Salt
1 Tbsp	Powdered milk
1 ½ tsp	Bread machine yeast
2 Tbsp	Butter

Procedure

Pre-mix the dry ingredients (except the yeast) using a whisk or fork in a bowl. Heat the water briefly (20–30 seconds) in the microwave. Add the honey to the water and pour the water into the bread pan. Add the dry ingredients to the bread pan. Cut the butter into four to eights small chunks and distribute evenly around the dry ingredients. Make a small indentation in the center of the dry ingredients, and put the yeast there. Set the bread machine for whole wheat, light crust and activate.

When bread is done, remove from the bread machine (Do this with oven gloves because the bread pan is hot!), wait until it is cool enough to handle but still warm, then slice into thick chunks. Slather the chunks with enough fresh creamy butter or peanut butter to make a cardiologist wince, and serve steaming.

Conclusion

As a culture, on average, we spend a lot of money on bread, cheese, wine, and beer. We don't usually spend as much on vinegar, but the vinegar we buy is usually of decidedly inferior quality.

Self-sufficiency has a lot of benefits. The first is that it saves you money. If you eat bread or drink wine or beer, knowing how to make your own could save you hundreds of dollars a year. If you usually consume all three, it could save you thousands of dollars a year. Those dollars represent time that could be better invested elsewhere. Even if some of that time is spent in making the foods, that time is still at home where you have access to your family and your family has access to you. The dollars you save make you less dependent on the decisions of an employer that has no loyalty to you. The dollars you save can be used to pay off debts so you have greater freedom, to fund a child's education, or to assure a secure retirement.

The second benefit is in the quality of the food. In one of the chapters on bread I included a picture of the ingredient label for some typical bread. Your own bread won't contain ingredients with six syllables. Furthermore, you can control the quality of the ingredients you use, and the cleanliness of your own procedures. Because you have an actual stake in the quality of what you produce and your goal is your own family's health, you aren't going to skirt the rules just to be more efficient or make an extra one-tenth of a penny per loaf or bottle. You can make your own wine from apple trees in your backyard that you know for a fact have never been touched with an artificial pesticide rather than relying on the honesty of people you have never met. The single person you can trust with your health more than anyone else is you.

The third benefit is that you can create unique items of superior quality or unique characteristics that simply cannot be purchased at any price. Examples would be my wine made from pears, apples, and honey or vinegar produced from such a wine. You will never find anything like it anywhere, and its quality is superior.

The final benefit is the most intangible and the most important. Knowledge is power. There is not only a certain pride that comes from taking control of your own destiny by producing your own food, but also a feeling of security. We live in an uncertain world. Terrorist attacks, regulations, world markets, tidal waves, earthquakes, and melted nuclear reactor cores can affect us directly and affect our food supply. Most items that we eat have traveled at least 1,000 miles in tractor trailers powered by petroleum, and the price of petroleum could explode any time due to the volatility of events in oil-producing regions.

With the information in this book and a little bit of practice to build confidence, you could literally start your own winery or bakery. If you grew barley, you could start your own brewery. You could make your own cheese with milk from a neighbor's cow, and trade that cheese to another neighbor for a hammer. This is the sort of knowledge that can be applied every day to simply improve your finances and the quality of your food, but also gives you confidence that you can provide for your family no matter what. It's the most valuable aspect of self-sufficiency, and the reason why I wrote this book. I hope you have enjoyed reading the book and applying the knowledge as much as I have enjoyed sharing it.

Brett Markham
New Ipswich, New Hampshire
2012

Index

S

Secondary fermentation, 23–25, 71–72, 97–98, 104, 120–122, 149–150
Sherry wines, 82
Solera aging, 84–85
Solutions, 8
Specific gravity, 11–12, 41, 43, 69
  of beer, 119–120
  weight method of determining, 39–40
Structural formulas, 9–10, 15
Sugar, 22, 38, 217
  adjusting, 40–42
  measuring, 38–40
  sources of, 64–65
Sulfites, 32, 53–54, 105
  testing, 54–55, 78–80

T

Tannins, 12–13, 30–31, 43–45, 75, 170
Tartaric acid, 30, 102, 183
Thermometer, 99, 182
Thiamine, 32, 53, 105
Titration, 47

V

Vinegar, 159, 183
  aging, 170–171
  conversion of ethanol to, 161
  dressing, 172
  equipment for making cheesecloth, 166–167
    rubber bands, 166–167
    vinegar crock, 165–166
    vinegar raft, 167
  herbal, 171–172
  in canning, 162
  making, 168–170
Vitis vinifera grapes, 22–23, 32, 38

W

Wine
  acidity of, 46–47
    adjusting, 48–51
    measuring, 47–48
  pectins, 51–52
  sugar levels
    adjusting, 40–42
    measuring, 38–40
  sulfite, 53–54
    testing, 54–55, 78–80
  tannins, 43–45
Winemaking
  cleanliness and sanitation, 65–66
  equipment, 29–30
    airlocks, 25–26
    corkers, 26–27
    nylon straining bags, 28
    primary fermenter, 24
    racking tube, 26, 70–71
    secondary fermenter, 24–25
    wine bottles, 28–29
    wine thief, 27–28
  ingredients of, 30–35
    fruits, 60–61
    herbs, 63
    juices, 61–62
    sources of sugar, 64–65
    spices, 63–64
    vegetables, 62–63
  process of, 37–38
  recipes for, 66–68, 73–74
  techniques, 23
    bottling wine, 72–73
    fortification, 81–83
    malolactic fermentation, 80–81
    oak aging, 83–84
    primary fermentation, 69–70
    racking, 70–71
  secondary fermentation, 71–72
  solera aging, 84–85
  without sulfites, 77–78

Y

Yeast, 22, 32, 55–56, 74, 78, 105–106, 137–138, 216–217
  Fermentis SafBrew S-33 Dry Ale, 106
  Lalvin D-47, 33
  Lalvin ICV-D254, 33–34
  Red Star Montrachet, 33
  Red Star Pasteur Champagne, 33
  Wyeast 1056 American Ale, 106–107
  Wyeast 1084 Irish Ale, 107
  Wyeast 4632 Dry Mead, 34
Yeast energizer, 32, 52–53, 105
Yeast nutrient, 52–53
Yogurt, 184–185, 190–191